# METHODS IN MOLECULAR BIOLOGY

*Series Editor*
**John M. Walker**
**School of Life and Medical Sciences**
**University of Hertfordshire**
**Hatfield, Hertfordshire, AL10 9AB, UK**

For further volumes:
http://www.springer.com/series/7651

# Computational Methods for Single-Cell Data Analysis

Edited by

## Guo-Cheng Yuan

*Dana–Farber Cancer Institute and Harvard Chan School of Public Health, Boston, MA, USA*

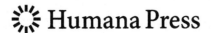 Humana Press

*Editor*
Guo-Cheng Yuan
Dana–Farber Cancer Institute and Harvard Chan
School of Public Health
Boston, MA, USA

ISSN 1064-3745          ISSN 1940-6029   (electronic)
Methods in Molecular Biology
ISBN 978-1-4939-9056-6      ISBN 978-1-4939-9057-3   (eBook)
https://doi.org/10.1007/978-1-4939-9057-3

Library of Congress Control Number: 2018967307

This Humana Press imprint is published by the registered company Springer Science+Business Media, LLC, part of Springer Nature.
The registered company address is: 233 Spring Street, New York, NY 10013, U.S.A.

# Preface

The cell is the fundamental unit of life. The biological functions of an organ or tissue are results of coordinated action of a large number of cells, each having its own properties and dynamic behavior. While it is traditional to classify cells with similar function and morphology as cell types, it is also well-recognized that, even within each cell type, there remain significant differences, or states, among individual cells. Current knowledge of the repertoire of cell types and cell states, as well as their dynamic changes, remains highly incomplete. Systematic, comprehensive characterization of spatial and temporal organization of cellular heterogeneity, along with the mechanisms underlying cell-type/state transition and maintenance, has important implications in development and diseases.

It is not until recently that it has become feasible to systematically investigate cellular heterogeneity at the single-cell resolution, thanks to the rapid development of a number of advanced technologies including sequencing, imaging, and microfluidic devices. Collectively, single-cell technologies have created exciting opportunities to systematically characterize the molecular behavior of individual cells at the omics scale. At the same time, the analysis and integration of single-cell omic data are difficult due to a number of challenges such as sparsity, technical variability, and spatial-temporal complexity.

During the past few years, numerous computational methods and software packages have been developed to overcome these challenges. The aim of this book is to introduce to the community the state of the art of computational approaches in single-cell data analysis. Each chapter presents a computational toolbox that is aimed to overcome a specific challenge in single-cell analysis, such as data normalization, rare cell-type identification, and spatial transcriptomics analysis. Rather than explaining the mathematical details, here the focus is on hands-on implementation of computational methods for analyzing experimental data. Taken together, these chapters cover a wide range of tasks and may serve as a handbook for single-cell data analysis.

Finally, I would like to thank Prof. John M. Walker for his kind invitation and sustained support throughout the preparation of this book. I would also like to express my sincere gratitude to all the contributors for sharing their protocols.

*Boston, MA, USA* *Guo-Cheng Yuan*

# Contents

# Contributors

MARTIN J. ARYEE • *Department of Biostatistics, Harvard T.H. Chan School of Public Health, Boston, MA, USA; Department of Pathology, Massachusetts General Hospital, Boston, MA, USA; Broad Institute of MIT and Harvard, Cambridge, MA, USA*

RHONDA BACHER • *Department of Biostatistics, University of Florida, Gainesville, FL, USA*

HAIDE CHEN • *Center for Stem Cell and Regenerative Medicine, Zhejiang University School of Medicine, Hangzhou, China; Stem Cell Institute, Zhejiang University, Hangzhou, China*

WEIYAN CHEN • *CAS Key Lab of Computational Biology, CAS-MPG Partner Institute for Computational Biology, Shanghai Institute of Nutrition and Health, Shanghai Institute of Biological Sciences, University of Chinese Academy of Sciences, Chinese Academy of Sciences, Shanghai, China*

MEICHEN DONG • *Department of Biostatistics, Gillings School of Global Public Health, University of North Carolina, Chapel Hill, NC, USA*

JEAN FAN • *Department of Chemistry and Chemical Biology, Harvard University, Boston, MA, USA*

LIJIANG FEI • *Center for Stem Cell and Regenerative Medicine, Zhejiang University School of Medicine, Hangzhou, China; Stem Cell Institute, Zhejiang University, Hangzhou, China*

GUOJI GUO • *Center for Stem Cell and Regenerative Medicine, Zhejiang University School of Medicine, Hangzhou, China; Stem Cell Institute, Zhejiang University, Hangzhou, China*

GARY C. HON • *Department of Obstetrics and Gynecology, Cecil H. and Ida Green Center for Reproductive Biology Sciences, University of Texas Southwestern Medical Center, Dallas, TX, USA*

YUANHUA HUANG • *EMBL-European Bioinformatics Institute, Cambridgeshire, UK*

HONGKAI JI • *Department of Biostatistics, Johns Hopkins Bloomberg School of Public Health, Baltimore, MD, USA*

ZHICHENG JI • *Department of Biostatistics, Johns Hopkins Bloomberg School of Public Health, Baltimore, MD, USA*

LAN JIANG • *Howard Hughes Medical Institute, Boston Children's Hospital, Boston, MA, USA; Program in Cellular and Molecular Medicine, Boston Children's Hospital, Boston, MA, USA; Division of Hematology/Oncology, Department of Pediatrics, Boston Children's Hospital, Boston, MA, USA*

PENG JIANG • *Regenerative Biology Laboratory, Morgridge Institute for Research, Madison, WI, USA*

YUCHAO JIANG • *Department of Biostatistics, Gillings School of Global Public Health, University of North Carolina, Chapel Hill, NC, USA; Department of Genetics, School of Medicine, University of North Carolina, Chapel Hill, NC, USA; Lineberger Comprehensive Cancer Center, University of North Carolina, Chapel Hill, NC, USA*

DIVY KANGEYAN • *Department of Biostatistics, Harvard T.H. Chan School of Public Health, Boston, MA, USA; Department of Pathology, Massachusetts General Hospital, Boston, MA, USA*

BEOMSEOK KIM • *Department of New Biology, DGIST, Daegu, Republic of Korea*

JONG KYOUNG KIM • *Department of New Biology, DGIST, Daegu, Republic of Korea*

ALICIA T. LAMERE • *Mathematics Department, Bryant University, Smithfield, RI, USA*

CALEB LAREAU • *Department of Biostatistics, Harvard T.H. Chan School of Public Health, Boston, MA, USA; Department of Pathology, Massachusetts General Hospital, Boston, MA, USA*

EUNMIN LEE • *Department of New Biology, DGIST, Daegu, Republic of Korea*

JUN LI • *Applied and Computational Mathematics and Statistics Department, University of Notre Dame, Notre Dame, IN, USA*

IDA LINDEMAN • *Wellcome Sanger Institute, Hinxton, Cambridge, UK; KG Jebsen Coeliac Disease Research Centre and Department of Immunology, University of Oslo, Oslo, Norway*

VILAS MENON • *Janelia Research Campus, Howard Hughes Medical Institute, Ashburn, VA, USA; Columbia University Medical Center, New York, NY, USA*

GUIDO SANGUINETTI • *School of Informatics, University of Edinburgh, Edinburgh, UK*

KARTHIK SHEKHAR • *Klarman Cell Observatory, Broad Institute of MIT and Harvard, Cambridge, MA, USA*

MICHAEL J. T. STUBBINGTON • *Wellcome Sanger Institute, Hinxton, Cambridge, UK*

HUIYU SUN • *Center for Stem Cell and Regenerative Medicine, Zhejiang University School of Medicine, Hangzhou, China; Stem Cell Institute, Zhejiang University, Hangzhou, China*

ANDREW E. TESCHENDORFF • *CAS Key Lab of Computational Biology, CAS-MPG Partner Institute for Computational Biology, Shanghai Institute of Nutrition and Health, Shanghai Institute of Biological Sciences, University of Chinese Academy of Sciences, Chinese Academy of Sciences, Shanghai, China; UCL Cancer Institute, University College London, London, UK*

SHIQI XIE • *Department of Obstetrics and Gynecology, Cecil H. and Ida Green Center for Reproductive Biology Sciences, University of Texas Southwestern Medical Center, Dallas, TX, USA*

YINCONG ZHOU • *Stem Cell Institute, Zhejiang University, Hangzhou, China; College of Life Sciences, Zhejiang University, Hangzhou, China*

QIAN ZHU • *Dana-Farber Cancer Institute, Boston, MA, USA*

# Chapter 1

## Quality Control of Single-Cell RNA-seq

### Peng Jiang

### Abstract

Single-cell RNA-seq (scRNA-seq) is emerging as a promising technology to characterize and dissect the cell-to-cell variability. However, the mixture of technical noise and intrinsic biological variability makes separating technical artifacts from real biological variation cells particularly challenging. Proper detection and filtering out technical artifacts before downstream analysis are critical. Here, we present a protocol that integrates both gene expression patterns and data quality to detect technical artifacts in scRNA-seq samples.

Key words  scRNA-seq, Quality control, Integrate, Gene expression patterns, Data quality

## 1  Introduction

Single-cell RNA-seq (scRNA-seq) provides a relatively unbiased approach to investigate the heterogeneity of cells in complex mixtures [1]. It has revolutionized our capacity to understand the transcriptomic diversity of cellular states [2, 3], lineages [4], and diseases [5]. However, one of the major challenges of this technology is the noise behind the data [6, 7]. For example, profiling low amounts of mRNA will likely lead to missing transcripts ("dropout" events) during the reverse-transcription step and also substantially distort original transcript abundance [6, 8]. Comparison of the top differentially expressed genes between cell populations shows poor consistency, suggesting that high variance could be caused by high-magnitude outliers [8]. On the other hand, gene expression among cells is inherently stochastic and the cell-to-cell variations can also be a result of transcriptional bursts or fluctuations [9]. Controlling quality of scRNA-seq and discarding technical artifacts are very important for downstream analysis.

To detect potential technical artifacts (bad samples) in scRNA-seq, previous studies have used various strategies that can be generally grouped into three categories. The first category involves using housekeeping genes to perform quality control (QC). For example, cells are filtered out if certain housekeeping genes (e.g., *Actb*,

*Gapdh*) are not expressed or abnormally expressed [10, 11]. The assumption of this approach is that housekeeping genes are highly and consistently expressed. This is true for bulk RNAs but it is not necessarily true for single cells (*see* **Note 1**). For instance, a study using single-cell qPCR has shown that the expressions of housekeeping genes have high variations among individual cells, and different cell types have distinguished housekeeping gene expression patterns [12]. Thus, a reliance on housekeeping genes to perform QC does not work for scRNA-seq samples. The second category for QC involves using overall gene expression patterns to define technical artifacts. For example, cells show distinguished gene expression patterns if compared with the majority of the cells are excluded from downstream analysis [13] (*see* **Notes 2–3**). The major problem of these methods in this category is that they may remove cells with real biological variation. The third category involves using the number of genes detected and/or the reads mapping rate to define technical artifacts [14]. However, the number of genes detected varies among experiments depending on the quality of a particular library, cell type, or RNA-protocol. The mapping rate cutoff is also hard to make, and thus the cutoff settings are typically arbitrary. Thus, although single-cell approaches hold great promise in investigating the cell heterogeneity, QC remains one of major challenges [7]. Nonetheless, our previous studies and own work demonstrate that integrating both gene expression patterns and sequencing data quality can be a reasonable strategy for performing QC [15]. The basic assumption of this approach is that if gene expression outliers are also associated with poor sequencing library quality, they are more likely to be technical artifacts than being real biological variation cells. We also assume that gene expression outliers contain both cells with real biological variation and technical artifacts, but the rest of the cells (main population cells) in general are more likely to contain good quality cells. Thus, we can use cells of the main population as controls to estimate data quality cutoffs and a corresponding false positive rate (FPR) (Fig. 1).

Here, we describe in detail our procedure for detecting technical artifacts in scRNA-seq using three batches of our published human embryonic stem cells (ES cells) scRNA-seq data [16].

## 2    Materials

### 2.1    Lab Equipment

1. C1 Single-Cell Auto Prep IFC (Fluidigm).

2. EVOS FL Auto Cell Imaging system (Life Technologies).

3. Illumina HiSeq 2500 system.

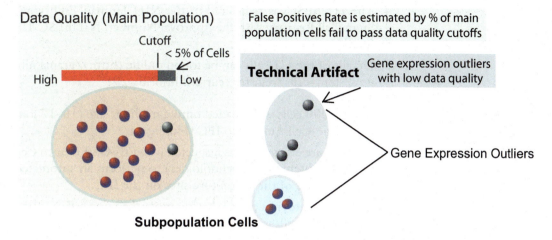

**Fig. 1** Illustration of quality control (QC) for scRNA-seq framework. Cells can be separated out based on gene expression patterns into gene expression outliers and cells of the main population. The data quality cutoffs are determined by allowing a certain percentage (e.g., <5%) of main population cells that fail to pass them. The technical artifacts are defined as gene expression outliers that fail to pass data quality cutoffs. The subpopulation cells are defined as gene expression outliers that can pass data quality cutoffs

**2.2 Kits**

1. SMARTer PCR cDNA Synthesis kit (Clontech).

2. Advantage 2 PCR kit (Clontech).

3. Nextera XT DNA Sample Preparation Index Kit (Illumina).

**2.3 ScRNA-seq Data**

1. Raw scRNA-seq dataset (H1) can be accessed by Gene Expression Omnibus (GEO) with accession number (GSE64016).

2. The downloaded files from GEO are SRA format.

3. SRA toolkit (http://www.ncbi.nlm.nih.gov/Traces/sra/sra.cgi?view=software) can be used to convert files from SRA format to FASTQ format via "fastq-dump" utility.

# 3 Methods

**3.1 H1 Human Embryonic Stem Cells (hESCs)**

1. Undifferentiated H1 human embryonic stem cells (hESCs) were cultured in E8 medium [17] on Matrigel-coated tissue culture plates with daily media feeding at 37 °C with 5% (vol/-vol) $CO_2$.

2. Cells were split every 3–4 days with 0.5 mM EDTA in $1 \times$ PBS for standard maintenance.

3. Immediately before preparing single-cell suspensions for each experiment, hESCs were individualized by Accutase (Life Technologies), washed once with E8 medium, and resuspended at densities of $5.0$–$8.0 \times 10^5$ cells/mL in E8 medium for cell capture.

4. The H1 hESCs is registered in the NIH Human Embryonic Stem Cell Registry with the Approval Number: NIHhESC-10-0043.

5. Details of the H1 cells can be found online (http://grants.nih.gov/stem_cells/registry/current.htm?id=29).

### 3.2 Single-Cell Capture and cDNA Library Preparation

1. 5000–8000 cells were loaded onto a medium size (10–17 $\mu m$) C1 Single-Cell Auto Prep IFC (Fluidigm).

2. The capture efficiency was inspected using EVOS FL Auto Cell Imaging system (Life Technologies) to perform an automated area scanning of the 96 capture sites on the IFC.

3. Empty capture sites or sites having more than one cell captured were first noted and those samples were later excluded from further library processing for RNA-seq.

4. Immediately after capture and imaging, reverse transcription and cDNA amplification were performed in the C1 system using the SMARTer PCR cDNA Synthesis kit (Clontech) and the Advantage 2 PCR kit (Clontech).

5. Full-length, single-cell cDNA libraries were harvested the next day from the C1 chip and diluted to a range of 0.1–0.3 ng/µL.

6. Diluted single-cell cDNA libraries were fragmented and amplified using the Nextera XT DNA Sample Preparation Kit and the Nextera XT DNA Sample Preparation Index Kit (Illumina).

7. Libraries were multiplexed at 24 libraries per lane, and single-end reads of 67-bp were sequenced on an Illumina HiSeq 2500 system.

### 3.3 Reads Mapping

1. Using Bowtie [18] to map raw reads against the reference genes (e.g., human hg19 Refseq reference) allowing up to two mismatches and a maximum of 20 multiple hits.

2. The mapped expected read counts and TPMs can be estimated by RSEM [19].

### 3.4 Classification of Cells into Gene Expression Outliers and Cells of the Main Population

1. Given a cell, calculate a list of Spearman rank correlations comparing that given cell to the rest of the cells in the dataset ("one-to-others").

2. Then, that given cell is removed and a list of pairwise Spearman rank correlations is calculated for the remaining cells ("pairwise").

3. Uses a one-sided Wilcoxon signed-rank test to assess whether the "one-to-others" correlation is significantly lower than the set of "pairwise" correlations.

4. A similar procedure is also performed using Pearson product-moment correlations.

5. Classify cells as either gene expression outliers or cells of the main population based on $p$-values of both tests.

6. In this study, we define gene expression outliers as cells with $p$-values less than 0.001 in both Spearman and Pearson tests.

### 3.5 Metrics to Evaluate the scRNA-seq Library Quality

1. Total number of mapped reads: the sum of mapped reads for all the genes. An extremely low number of mapped reads may affect the ability to characterize the transcriptome and could be due to either a low mapping rate or other technical issues introduced during sample prep or sequencing.

2. Mapping rate: the total number of mapped reads divided by the read depth. Mapping rate can be effected by RNA degradation, contamination with genomic DNA, or other technical issues introduced during sample prep or sequencing.

3. Reads complexity: the ratio of unique reads (the count of reads after removing duplicates) over the total number of all reads.

### 3.6 Combining Library Quality Metrics to Combined Scores

1. For each cell, calculates a quantile score (QS) for each quality metric. Given a metric, the QS of a cell is defined as the number of other cells in the dataset with equal or lower values divided by the total number of cells. For example, if a cell has the 20th highest mapping rate among a set of 80 cells, then the mapping rate QS for this particular cell is 0.75. A higher QS indicates better data quality.

2. Minimal Quantile Score (MQS): the minimal QS of the three quality metrics.

$$MQS = min\{QS_i\}$$

$$i \in \{mapped\ reads, mapping\ rate, reads\ complexity\}$$

MQS assumes that each of the three quality metrics is critical and that a deficiency in any of the three is a potential indicator of technical issues. Thus the "final quality" of a cell depends on its lowest quality metric score.

3. Weighted Combined Quality Score (WCQS): WCQS assumes that the importance of each quality metric may depend on specific experimental batches, protocols, and/or conditions. WCQS assumes that the importance of each quality metric for detecting technical artifacts is proportional to its ability to discriminate between gene expression outliers and cells of the main population. For example, given a batch of cells, if the mapping rate of a given batch of cells can perfectly discriminate between gene expression outliers and cells of the main population, then it is more likely that the mapping rate is a dominant player in detecting technical artifacts. In contrast, if a metric does not indicate differences between gene expression outliers and cells of the main population, then it should be removed

from prediction of potential technical artifacts. WCQS calculates a weighted aggregation quality score for each sample defined as:

$$\text{WCQS} = \frac{\sum_{i=1}^{k} w_i Z_i}{\sqrt{\sum_{i=1}^{k} w_i^2}}$$

where $Z_i$ is the transformed $Z$-score of QS for data quality metric $i$, according to $Z_i = \Phi^{-1}(1 - P_i)$ where $\Phi$ is the standard normal cumulative distribution function, and $P_i$ is the probability that quality metric $i$ is lower in a given cell than in the rest of the cells. We estimate $P_i$ is as $P_i = 1 - \text{QS}_i$. To avoid numerical error, we can set maximal and minimal $Z_i$ as $+8.5$ and $-8.5$, which corresponds to $P_i < 10^{-16}$ and $P_i > (1-10^{-16})$, respectively. The $w_i$ is the weighting factor for data quality metric $i$ and is estimated according to the individual quality metric's ability to discriminate between cells of the main population and gene expression outliers as

$$w_i = \begin{cases} \dfrac{\text{AUC}_i - 0.5}{0.5} & (\text{AUC}_i > 0.5) \\ 0 & (\text{AUC}_i \leq 0.5) \end{cases}$$

where $\text{AUC}_i$ is the area under the curve (AUC) of the receiver operating characteristic (ROC) curve for quality metric $i$. If a quality metric $i$ (e.g., mapping rate) can perfectly discriminate between cells of the main population and gene expression outliers, then $\text{AUC}_i = 1$ and thus $w_i = 1$. If the values of a quality metric $i$ are randomly distributed in cells of the main population and gene expression outliers, then the expected $\text{AUC}_i = 0.5$ and thus $w_i = 0$.

*3.7  Identification of Technical*

1. We assume that good quality cells should pass particular MQS and WCQS cutoffs. We use cells of the main population as controls to determine these cutoffs (*see* **Note 4**). You can enumerate all possible combinatorial pairs of MQS and WCQS cutoffs in a given dataset, calculate the fraction of cells of the main population that pass both cutoffs of a pair, and then uses the remaining cells of the main population to estimate the corresponding false positive rate (FPR) for that pair (Fig.1).

2. If more than one pair of MQS and WCQS cutoffs results in the same FPR, you can choose the cutoff pair that maximizes the percentage of gene expression outliers failing to pass.

3. Applies these cutoffs to the gene expression outliers to identify technical artifacts. Technical artifacts are defined as gene expression outliers with poor data quality measurements.

*3.8  SinQC Software*

1. SinQC [15] is designed for implementing (Subheadings 3.3–3.6) (*see* **Note 5**).

2. The SinQC software and detailed user manual are available at http://www.morgridge.net/SinQC.html

# 4    Notes

*[handwritten margin note: can't use, housekeeping genes]*

1. Several studies used housekeeping genes to perform quality control for scRNA-seq datasets [10, 11]. To further investigate the feasibility of using housekeeping genes to perform quality control for scRNA-seq datasets, we calculated the gene expression levels (TPMs) for two housekeeping genes (*Actb* and *Gapdh*) in a mouse scRNA-seq dataset [20]. The *Gapdh* is significantly higher expressed in ES cells than in MEF cells ($P = 5.6e{-}06$, 1-sided Wilcoxon rank sum test) while the *Actb* is significantly lower expressed in ES cells than in MEF cells ($P < 2.2e{-}16$, 1-sided Wilcoxon rank sum test) [15]. It suggests that it is infeasible to use housekeeping genes to perform QC for scRNA-seq dataset.

2. Using median gene expression values or the number of genes detected (TPM > 1) to perform quality control (QC): Low data quality (e.g., low mapping rate) can result in fewer number of genes detected or low median gene expression values. However, the number of genes detected (TPM > 1) can also be biologically related. The number of genes detected varies depending on the quality of a particular library and cell types [8]. We calculated the number of genes detected in a highly heterogeneous scRNA-seq dataset containing 301 cells (mixture of 11 different cell types) [4]. The number of genes detected is highly cell type dependent, suggesting using the number of gene detected to identify technical artifacts will result in substantial bias ([15], Fig. S8). For highly heterogeneous scRNA-seq datasets, the technical artifacts detected by this method are more likely to have fewer genes detected if compared with QC pass cells. But this does not mean that the cells with fewer genes detected are technical artifacts.

3. Using "genes detected and/or mapping rate" to perform quality control (QC): The basic idea of using "genes detected and/or mapping rate" [14] to perform QC is that the fewer number of genes detected could be due to both technical issues and biological heterogeneity. But if a cell with fewer genes detected is also associated with low mapping rate (mapping rate is technical related), the cell might be more likely to be a technical artifact. This approach is the most conceptually similar to our method. However, our method has strengths on two aspects: First, since the mapping rate and the number of genes detected are not directly correlated, the mapping rate cutoff and the number of genes detected cutoff chosen are very difficult and arbitrary. Our method maximizes the probability that the technical artifacts are correctly detected while also minimizing the false positives by using cells of the main population as data quality controls. Second, in addition to mapping rate, our method also takes other library quality metrics into consideration (e.g., library complexity).

4. Our method assumes that gene expression outliers contain both technical artifacts and biological variant cells, but cells of the main population, in general, are more likely to contain good quality cells. Thus, our method uses cells of the main population as controls to estimate data quality score cutoffs and a corresponding false positive rate (FPR). However, given a FPR, it is a challenge to estimate the corresponding false negative rate (technical artifacts that are missed), due to that scRNA-seq has no "ground-truth" for "bad samples." Sensitivity (also called the true positive rate) is the proportion of positives ("technical artifacts") that are correctly identified. Specificity (also called the true negative rate) measures the proportion of negatives ("good quality single cells") that are correctly identified. Since scRNA-seq has no "ground-truth" for "good samples" and "bad samples," it is a challenge to estimate these two measurements directly. To further compare the sensitivity and specificity of our method in high-heterogeneity and low-heterogeneity datasets, we applied our method to datasets with mixture of different portions of cell types, and compared the overlap of technical artifacts detected among them. For example, using a mouse scRNA-seq dataset (48 ES cells and 44 MEF cells) [20], we mixed the cells into three different categories: high-heterogeneity (48 ES cells +44 MEF cells), medium-heterogeneity ("ES cells (all) + 1/5 (MEF) cells" and ("MEF cells (all) + 1/5 (ES) cells"), and low heterogeneity ((48 ES cells) and (44 MEF cells), separately). Our method detects two technical artifacts (ESC_46 and ESC_32) in the high-heterogeneity dataset (48 ES cells +44 MEF cells). These two technical artifacts can also be robustly detected either in medium-heterogeneity dataset or low heterogeneity dataset. However, if we apply our method to each individual ES (48 cells) or MEF (44 cells) dataset separately, we can detect more artifacts, comparing to apply our method on pooled mixture datasets (48 ES cells +44 MEF cells). Therefore, we conclude that our method increases specificity at the cost of dropping sensitivity when the extent of heterogeneity in a dataset is high. In highly heterogeneous cell populations, detecting technical artifacts carries a higher risk of dropping real biological variation cells. The increased specificity and decreased sensitivity of our method for highly heterogeneous cell populations is a good feature that can minimize the false positives.

5. The running SinQC for scRNA-seq QC is not restrictive to RSEM output files ("*.genes.results"). For users who do not use RSEM, they can make a customized RSEM files ("*.genes.results") to run SinQC. A detailed manual can be found in SinQC website (http://www.morgridge.net/SinQC.html).

## References

1. Eberwine J, Sul J-Y, Bartfai T, Kim J (2014) The promise of single-cell sequencing. Nat Methods 11(1):25–27

2. Trapnell C, Cacchiarelli D, Grimsby J, Pokharel P, Li S, Morse M et al (2014) The dynamics and regulators of cell fate decisions are revealed by pseudotemporal ordering of single cells. Nat Biotechnol 32(4):381–386. Epub 2014/03/25. https://doi.org/10.1038/nbt.2859

3. Xue Z, Huang K, Cai C, Cai L, Jiang CY, Feng Y et al (2013) Genetic programs in human and mouse early embryos revealed by single-cell RNA sequencing. Nature 500(7464):593–597. Epub 2013/07/31. https://doi.org/10.1038/nature12364

4. Pollen AA, Nowakowski TJ, Shuga J, Wang X, Leyrat AA, Lui JH et al (2014) Low-coverage single-cell mRNA sequencing reveals cellular heterogeneity and activated signaling pathways in developing cerebral cortex. Nat Biotechnol 32(10):1053–1058. Epub 2014/08/05. https://doi.org/10.1038/nbt.2967

5. Patel AP, Tirosh I, Trombetta JJ, Shalek AK, Gillespie SM, Wakimoto H et al (2014) Single-cell RNA-seq highlights intratumoral heterogeneity in primary glioblastoma. Science 344 (6190):1396–1401. Epub 2014/06/14. https://doi.org/10.1126/science.1254257

6. Sandberg R (2014) Entering the era of single-cell transcriptomics in biology and medicine. Nat Methods 11(1):22–24

7. Stegle O, Teichmann SA, Marioni JC (2015) Computational and analytical challenges in single-cell transcriptomics. Nat Rev Genet 16 (3):133–145

8. Kharchenko PV, Silberstein L, Scadden DT (2014) Bayesian approach to single-cell differential expression analysis. Nat Methods 11 (7):740–742

9. Munsky B, Neuert G, van Oudenaarden A (2012) Using gene expression noise to understand gene regulation. Science 336 (6078):183–187

10. Ting DT, Wittner BS, Ligorio M, Vincent Jordan N, Shah AM, Miyamoto DT et al (2014) Single-cell RNA sequencing identifies extracellular matrix gene expression by pancreatic circulating tumor cells. Cell Rep 8 (6):1905–1918. Epub 2014/09/23. https://doi.org/10.1016/j.celrep.2014.08.029

11. Treutlein B, Brownfield DG, Wu AR, Neff NF, Mantalas GL, Espinoza FH et al (2014) Reconstructing lineage hierarchies of the distal lung epithelium using single-cell RNA-seq. Nature 509(7500):371–375. Epub 2014/04/18. https://doi.org/10.1038/nature13173

12. Oyolu C, Zakharia F, Baker J (2012) Distinguishing human cell types based on housekeeping gene signatures. Stem Cells 30 (3):580–584

13. Zeisel A, Muñoz-Manchado AB, Codeluppi S, Lönnerberg P, La Manno G, Juréus A et al (2015) Cell types in the mouse cortex and hippocampus revealed by single-cell RNA-seq. Science 347(6226):1138–1142

14. Kumar RM, Cahan P, Shalek AK, Satija R, DaleyKeyser AJ, Li H et al (2014) Deconstructing transcriptional heterogeneity in pluripotent stem cells. Nature 516(7529):56–61. Epub 2014/12/05. https://doi.org/10.1038/nature13920

15. Jiang P, Thomson JA, Stewart R (2016) Quality control of single-cell RNA-seq by SinQC. Bioinformatics 32(16):2514–2516. https://doi.org/10.1093/bioinformatics/btw176

16. Leng N, Chu LF, Barry C, Li Y, Choi J, Li X, et al (2015) Oscope identifies oscillatory genes in unsynchronized single-cell RNA-seq experiments. Nat Methods 12(10):947–950. https://doi.org/10.1038/nmeth.3549. PubMed PMID: 26301841; PubMed Central PMCID: PMC4589503

17. Chen G, Gulbranson DR, Hou Z, Bolin JM, Ruotti V, Probasco MD et al (2011) Chemically defined conditions for human iPSC derivation and culture. Nat Methods 8(5):424–429. https://doi.org/10.1038/nmeth.1593

18. Langmead B, Trapnell C, Pop M, Salzberg SL (2009) Ultrafast and memory-efficient alignment of short DNA sequences to the human genome. Genome Biol 10(3):R25. https://doi.org/10.1186/gb-2009-10-3-r25

19. Li B, Dewey CN (2011) RSEM: accurate transcript quantification from RNA-Seq data with or without a reference genome. BMC Bioinformatics 12:323. https://doi.org/10.1186/1471-2105-12-323

20. Islam S, Kjällquist U, Moliner A, Zajac P, Fan J-B, Lönnerberg P et al (2011) Characterization of the single-cell transcriptional landscape by highly multiplex RNA-seq. Genome Res 21 (7):1160–1167

# Chapter 2

# Normalization for Single-Cell RNA-Seq Data Analysis

Rhonda Bacher

## Abstract

In this chapter, we describe a robust normalization method for single-cell RNA sequencing data. The procedure, SCnorm, is implemented in R and is part of Bioconductor. Also included in the package are diagnostic functions to visualize normalization performance. This chapter provides an overview of the methodology and provides example work-flows.

**Key words** Single-cell RNA-seq, Normalization, Gene expression, Read count, High-throughput sequencing

## 1   Introduction

The purpose of *normalization* is to remove the effects of systematic technical variability on the observed gene expression measurements. Popular RNA-seq analyses such as sample clustering or classification, and differential gene expression require between-sample normalization as a first step to ensure measurements are comparable across samples. Variation in *sequencing depth* (the total amount each sample has been sequenced) is one particular artifact that arises during the RNA-seq experiment. As samples are sequenced more deeply, all gene counts should, on average, increase proportionally [1]. We call this dependence of the read counts on sequencing depth the *count-depth relationship* [2].

Most existing normalization methods developed for bulk RNA-seq experiments calculate global scale factors to adjust each sample for sequencing depth (one scale factor per sample is applied to all genes in the sample) [1, 3–5]. These methods work well when the count-depth relationship is common across genes, as it is in bulk RNA-seq data (Fig. 1a). When global scale-factor normalization [3] is effective, the resulting normalized count-depth relationships are near zero for most genes (Fig. 1b). Conversely, in single-cell RNA-seq data (scRNA-seq) genes exhibit variability in the count-depth relationship (Fig. 1c). In this case, when the count-

Guo-Cheng Yuan (ed.), *Computational Methods for Single-Cell Data Analysis*, Methods in Molecular Biology, vol. 1935, https://doi.org/10.1007/978-1-4939-9057-3_2, © Springer Science+Business Media, LLC, part of Springer Nature 2019

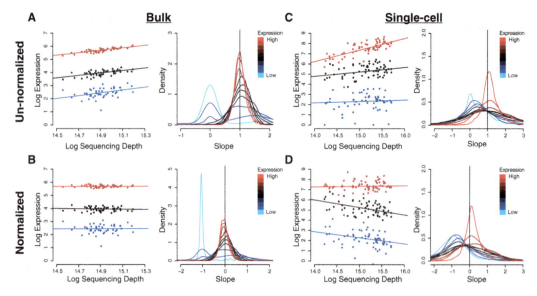

**Fig. 1** For each gene, median quantile regression was used to estimate the count–depth relationship for a bulk and single-cell RNA-seq dataset before and after normalization. (**a**) The left plot shows log un-normalized expression versus log depth with estimated regression fits for three genes in a bulk RNA-seq dataset containing no zero measurements and having low, moderate, and high expression defined as the median expression among nonzero un-normalized measurements in the 10th to 20th quantile (blue), 40th to 50th quantile (black), and 80th to 90th quantile (red), respectively. On the right, the estimated regression fits for all genes within ten equally sized gene groups where genes were grouped by their median expression among nonzero un-normalized measurements. (**b**) Similar to **a** for normalized bulk RNA-seq data. (**c** and **d**) Similar to **a** and **b** for single-cell RNA-seq data

depth relationship is not common across genes, the application of traditional global scale factor normalization methods leads to biased counts (Fig. 1d).

Our method, SCnorm [2], addresses the variability in the count-depth relationship in scRNA-seq data. SCnorm uses quantile regression to estimate the count-depth relationship for each gene. Genes with similar relationships are grouped together, and a second quantile regression estimates scaling factors within each group. In this chapter, we will further describe the SCnorm method and its software implementation.

## 2   Materials

Our method for normalizing single-cell RNA-seq data is implemented in the R package SCnorm and is available on Bioconductor:

https://bioconductor.org/packages/devel/bioc/html/SCnorm.html

The package is compatible with R versions greater than 3.4.0. The package may be installed directly from Bioconductor:

```
source("https://bioconductor.org/biocLite.R")
biocLite("SCnorm")
```

## 3    Methods

We describe our approach step-by-step as follows:

1. *Filter genes.* Filtering of genes with very low average expression is a common preprocessing step in RNA-seq analyses. As detailed in **Note 1**, we require genes to have at least ten nonzero expression counts. Filtered genes will not be included in the normalization.

2. *Estimate count-depth relationship per gene.* Let $\Upsilon_{g,j}$ denote the log nonzero expression for gene $g$ in cell $j$ for $g = 1,...,m$ and $j = 1,...,n$. Let $X_j$ denote the log sequencing depth, calculated as $\log\left(\sum_{g=1}^{m} e^{\Upsilon_{g,j}}\right)$. A median quantile regression is fit to estimate each gene's relationship between log un-normalized expression and log sequencing depth as

$$Q^{0.5}\left(\Upsilon_{g,j}|X_j\right) = \beta_{g,0} + \beta_{g,1}X_j$$

The estimated slope, $\widehat{\beta}_{g,1}$, represents the count-depth relationship for gene $g$.

3. *Group genes.* Genes are clustered into $K$ groups based on their $\widehat{\beta}_{g,1}$ using the $K$-medoids algorithm [6] via the `clara` function in the R package `cluster` [7]. If a cluster contains less than 100 genes, it is joined to the nearest cluster based on the cluster centers (medoids). Starting at $K = 1$, SCnorm will increase $K$ to $K + 1$ until a convergence criterion is satisfied (described below in **step 5.** *Evaluating K*).

4. *Group fit.* Within a homogenous gene group, a representative gene could be selected and the scale factors calculated as the ratio of its counts in each sample to the gene's overall mean. However, due to numerous zeros in the data and high levels of variability [8], a representative gene for the group must be estimated. We estimate a representative gene as the predicted values from a quantile regression between the log nonzero un-normalized expression counts from all genes in the group and log sequencing-depth. For computational reasons, we restrict the number of genes considered (*see* **Note 1**) to the 25% whose $\widehat{\beta}_{g,1}$ is closest to the group's overall count-depth relationship, estimated as $^{\mathrm{mode_g}}\widehat{\beta}_{g,1}$.

When fitting the quantile regression, the median may not always best represent subtle effects of the count-depth relationship for genes in a particular group, therefore we consider multiple quantiles $\tau$ and degrees $d$ and fit:

$$Q^{\tau_k, d_k}\left(\Upsilon_j | X_j\right) = \beta_0^{\tau_k} + \beta_1^{\tau_k} X_j + \ldots + \beta_d^{\tau_k} X_j^{d_k}$$

We fit over a grid of combinations of $\tau$ and $d$ from $\tau \in (.05, .10,$ $\ldots, .85, .90)$ and $d \in (1, 2, \ldots, 6)$. For each model fit, the predicted values, $\widehat{\Upsilon}_j^{\tau_k, d_k}$, can be viewed as a representative gene. To evaluate which specific values of $\tau_k$ and $d_k$, $\tau_k^*$ and $d_k^*$, are optimal, we choose the model in which the predicted gene's count-depth relationship is closest to the group's mode. Specifically, the predicted gene's count depth relationship, $\widehat{\eta}_1^{\tau_k, d_k}$, is estimated using median quantile regression:

$$Q^{0.5}\left(\widehat{\Upsilon}_j^{\tau_k, d_k} | X_j\right) = \eta_0^{\tau_k, d_k} + \eta_1^{\tau_k, d_k} X_j$$

The selected model is then the $\tau_k$ and $d_k$ which minimize $|\widehat{\eta}_1^{\tau_k, d_k} - {}^{\text{mode}_g}\widehat{\beta}_{g,1}|$

Once the best model is selected, the scale factors are estimated as $\text{SF}_{j,k} = \dfrac{e^{\widehat{\Upsilon}_j^{\tau_k^*, d_k^*}}}{e^{\Upsilon^{\tau_k^*}}}$ where $\Upsilon^{\tau_k^*}$ is the $\tau^{*\text{th}}$ quantile of expression counts in the $k^{\text{th}}$ group. The normalized counts $\Upsilon'_{g,j}$ are given by $\dfrac{e^{\Upsilon_{g,j}}}{\text{SF}_{j,k}}$.

*Evaluating K.* To determine if a given $K$ is sufficient, we evaluate the count-depth relationship of the normalized data. For every gene, its log normalized counts are regressed against the original log sequencing depth in a median quantile regression:

$$Q^{0.5}\left(\Upsilon'_{g,j} | X_j\right) = \beta'_{g,0} + \beta'_{g,1} X_j$$

We check that the distribution of $\beta'_{g,1}$ is centered around zero for genes across all expression levels. To do so, we group genes into ten equally sized groups based on their median nonzero expression. For each group, the mode of the slopes is estimated. Any mode outside of $(-0.1, 0.1)$ is evidence that the current $K$ is not optimal, and in which case $K$ is increased by one and the group fit and evaluation steps are repeated (*see* **Note 2** for an example).

*Multiple Conditions.* When multiple conditions or batches are present, SCnorm utilizes the procedure above for each condition and then rescales across all cells. During rescaling, all genes are split into quartiles based on their nonzero median un-normalized expression measurements. Within each quartile and condition, each gene is adjusted by a common scale factor defined as the median of the gene specific fold-changes, where fold-changes are calculated between each gene's condition-specific mean and its mean across conditions. Means are calculated over nonzero counts (*see* **Note 3** for additional details). If spike-ins are available and are representative of the biological genes, they may be used in this step to estimate the across condition scaling factors (additional details are in **Note 4**).

# 4  Notes

Note 1 reviews all arguments for running the main functions in the SCnorm package. Note 2 demonstrates a standard workflow for normalizing scRNA-seq data and evaluating the normalization. Note 3 reviews important considerations for normalization when multiple conditions are present. Note 4 details how to use spike-ins for across condition scaling.

1. There are two main functions accessible by the user in the SCnorm package:

   SCnorm—implements the normalization procedure.

   plotCountDepth—estimates and graphically displays the count-depth relationships.

   The SCnorm function requires only two arguments: the un-normalized expression matrix (Data) and a vector denoting the condition or batch each cell belongs to (Conditions). The Data argument should contain un-normalized expression measurements with genes (or other features) on the rows and cells on the columns. The expected format of Data is a data matrix in R or of class SummarizedExperiment of the SummarizedExperiment R package [9]. The Conditions argument should be a vector with length equal to the number of cells and match the exact order of the columns of the Data argument.

   The SCnorm function will implement the entire normalization procedure described above, automatically iterating until an optimal $K$ is reached. A dataset with 100 cells and 20,000 genes will take approximately 15 min to run with three computing cores. The computation time will increase as the number of cells and genes increases, though increasing the number of cores can be used to offset the increased time. In the example given, increasing to seven cores reduces the time to around 8 min.

   The output of SCnorm is a SummarizedExperiment object containing at minimum the normalized data (NormalizedData) and a list of the genes not included in the normalization (Genes-FilteredOut). Additional outputs may be generated using the non-default options and are described in more detail below.

   The full SCnorm function with default arguments is:

```
SCnorm(Data = NULL, Conditions = NULL, PrintProgressPlots = FALSE,
       reportSF = FALSE, FilterCellNum = 10, FilterExpression = 0,
       Thresh = 0.1, K = NULL, NCores = NULL, ditherCounts = FALSE,
       PropToUse = 0.25, Tau = 0.5, withinSample = NULL, useZerosToScale
= FALSE,
       useSpikes = FALSE)
```

Additional options the user may specify are:

`PrintProgressPlots`: If set to `TRUE`, SCnorm will automatically produce count-depth plots evaluating each value of $K$ attempted. It is highly recommended to set this option to `TRUE` and wrap the SCnorm function call in a pdf (or other graphics) device. Although SCnorm determines the optimal $K$ internally, viewing these plots is one way to evaluate the results of the normalization.

`reportSF`: If set to `TRUE`, SCnorm will output the scaling factors calculated for each group. This may be useful if users plan to conduct downstream differential expression (DE) tests, where the matrix of scale factors may be provided to functions in the EBSeq [10] and DESeq2 [11] packages which require the un-normalized data as input. For scRNA-seq DE tools, such as MAST [12] and scDD [13], the normalized counts may be used directly.

`FilterExpression`: A single numeric value denoting the cutoff used to exclude genes from the normalization based on average expression. Occasionally (but rarely) very lowly expressed genes with many tied counts may hinder the convergence of SCnorm and need to be filtered from the normalization procedure.

`FilterCellNum`: The minimum number of nonzero counts required for a gene to be included in the normalization. This value must be larger than or equal to ten. Since SCnorm fits a median quantile regression for each gene, we require a minimum of ten nonzero expression values for each gene by default. This is usually sufficient in practice although the user may wish to further filter genes that have expression in at least some proportion of cells. In that case, the proportion should be converted to an integer for the specific experiment to be normalized.

`Tau`: This option likely does not need to be adjusted by the user. It specifies the quantile used to estimate the quantile regression. We recommend using the median (`Tau=.5`) quantile.

`PropToUse`: During the group fitting, only a subset of genes is used in order to speed up computation time. We recommend using 25% of genes in the group. However, if the data are summarized by transcripts rather than genes, additional reduction in computation time may be obtained by setting this to 10%.

`Thresh`: During the evaluation procedure, SCnorm will calculate the distance of the normalized slope modes from zero. The default distance is 0.1 and in practice this offers the best trade-off between bias and computation time. A smaller value will take longer to converge, while a larger value may not result in the optimal normalization.

`K`: The number of groups to split the genes into for normalization. If any numeric value is supplied the iterative procedure of SCnorm will not be performed. Users are not advised to set this option. However, it may be helpful in quickly obtaining normalized data in the instance that the user previously ran the normalization but did not save the data or wishes to change arguments that only affect the across-condition scaling step.

NCores: The number of cores to use. The more cores available, the faster SCnorm will perform. By default, SCnorm will use one less than the number of cores available on the machine.

ditherCounts: When this option is set to TRUE, counts will be randomly jittered by 0.01 prior to fitting. With unique molecular identifier (UMI) scRNA-seq experiments, the data typically have many tied count values, which occasionally cause the quantile regression fit to fail. We find that dithering the counts by a small value avoids this issue and does not otherwise affect the normalization procedure or resulting normalized counts.

withinSample: As demonstrated in previous papers, gene-specific features may vary across samples. We have implemented the method from Risso et al. [14] if the user wishes to first normalize the counts based on a gene-specific feature such as GC content or gene length. This argument expects a vector of equal length to the number of rows of Data (and in matching order) with values representing the gene-specific feature to normalize. Note that within sample normalization should be used with caution as it is often specific to the experiment and exploratory analyses are highly recommended.

useZerosToScale: If set to TRUE, the zeros will be used when scaling across conditions. Use of this argument depends on which downstream differential expression tool will be used. If using methods which test zeros separately from continuous counts, such as MAST [12] or scDD [13], this option should remain FALSE. However, for methods such as DESeq2 [11] which test all counts together, this flag should be set to TRUE. A detailed example is given in **Note 3.**

useSpikes: We do not implement the use of spike-ins for within group normalization at this time because there are currently too few to estimate scale factors robustly in all groups. However, when multiple conditions or batches are being normalized, if this argument is TRUE then spike-ins will be used to perform the across condition scaling. The spike-ins are expected to be named following the convention of "ERCC-". Additional details regarding the use of spike-ins is given in **Note 4.**

The plotCountDepth function is used to visualize the count-depth relationships. It includes a wrapper for internal functions that estimate the count-depth relationships and then outputs a plot. During the normalization, if PrintProgressPlots=TRUE, multiple calls will be made to the plotCountDepth function, otherwise the function may be used stand-alone. The required arguments are Data and Conditions similar to the SCnorm function. All genes will be split into ten equally sized groups based on their nonzero un-normalized median expression. A density curve of the gene-specific count-depth relationships will be plot with a different color for each expression group. In addition to outputting a plot, the argument will provide a list object with each gene's name, the expression group it belongs to, and its estimated count-depth relationship.

The default arguments for the `plotCountDepth` function are:

```
plotCountDepth <- function(Data, NormalizedData=NULL, Conditions =
NULL,
              Tau = .5, FilterCellProportion = .10,
              FilterExpression = 0, NumExpressionGroups =
10,
              NCores=NULL, ditherCounts = FALSE)
```

Most of the additional arguments are the same as for the `SCnorm` function, including `ditherCounts`, `NCores`, `Filter-Expression`, and `Tau`. Here we use the option `FilterCellPro-portion`, which allows the user to directly filter genes from consideration if they do not have a certain proportion of cells expressed. Arguments unique to this function are:

`NumExpressionGroups`: The number of equally sized expression groups used to visualize the count-depth relationships. We typically use ten groups.

`NormalizedData`: If evaluating normalized data, then this argument specifies the normalized data matrix (the format is expected to be the same as the `Data` argument). Supplying this argument is critical to ensure the evaluation of the normalized measurements is done in terms of the original sequencing depths.

2. Here we apply SCnorm to a scRNA-seq dataset of 92 H1 human embryonic stem cells. Prior to sequencing, and follow-ing library preparation, the cDNA for each cell was split into two pools. One pool was sequenced with 96 cells per lane, while the other pool was sequenced with 24 cells per lane. Since lanes average similar numbers of total reads, this setup results in one pool of cells having an average sequencing depth of one million (H1-1 M) and the other with an average of four million (H1-4 M). Prior to normalization, the H1-4 M group will appear four times higher on average than the H1-1 M group. Since the cells are exactly the same, these data provide a benchmark to evaluate normalization procedures. It may be downloaded directly from GEO (file: GSE85917_Bacher. RSEM.xlsx):

https://www.ncbi.nlm.nih.gov/geo/query/acc.cgi? acc=GSE85917

We will use the first two sheets in the Excel file, which can be loaded into R by:

```
> library(readxl)
> h1cells.4M <-
data.frame(read_excel("GSE85917_Bacher.RSEM.xlsx",
                      sheet=1), stringsAsFactors=F)
> h1cells.1M <-
data.frame(read_excel("GSE85917_Bacher.RSEM.xlsx",
                      sheet=2), stringsAsFactors=F)
```

Next, we visualize the variability in the count-depth relationship across each of the datasets.

```
> library(SCnorm)
> cdr.1M <- plotCountDepth(Data = h1cells.1M, Conditions =
rep("1M",
                        ncol(h1cells.1M)))
> cdr.4M <- plotCountDepth(Data = h1cells.4M, Conditions =
rep("4M",
                        ncol(h1cells.4M)))
```

The results are shown in Fig. 2a, b. The count-depth relationship varies across genes, indicating that SCnorm should be used to normalize the data.

Using SCnorm we can normalize each batch independently and then scale across the batches by specifying the Conditions argument:

```
> Conditions <- c(rep("1M", ncol(h1cells.1M)), rep("4M",
                   ncol(h1cells.4M)))
> allData <- cbind(h1cells.1M, h1cells.4M)
> myNormData <- SCnorm(Data = allData, Conditions = Conditions,
                   PrintProgressPlots = TRUE, reportSF = TRUE)
```

Since `PrintProgressPlots=TRUE`, each attempted value of *K* will be plot until the procedure reaches convergence (Fig. 2c, d). SCnorm finished in 11 min using three cores.

The output is obtained by using the `results` function:

```
> normData <- results(myNormData, type = "NormalizedData")
> genesOUT <- results(myNormData, type = "GenesFilteredOut")
> scaleFactors <- results(myNormData, type = "ScaleFactors")
```

3. Single-cell specific differential expression methods such as MAST [12] and scDD [13] test the continuous (or nonzero) counts separately from zero counts. Thus, the default option

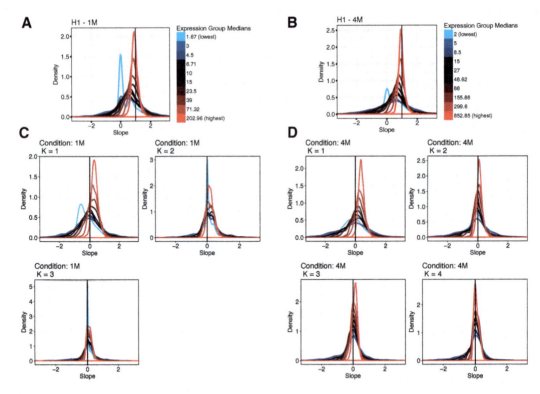

**Fig. 2** Count-depth relationships before and during normalization for the H1-1 M and H1-4 M data. (**a**) For the H1-1 M dataset, the estimated count-depth relationships for all genes within ten equally sized gene groups where genes were grouped by their median expression among nonzero un-normalized measurements. (**b**) Similar to **a** for the H1-4 M dataset. (**c**) The count-depth relationship is shown for the normalized counts for each value of *K* tried by SCnorm for the H1-1 M dataset. The genes remain in their initial expression groups as shown in **a**. (**d**) Similar to **C** but for the H1-4 M dataset

for SCnorm when scaling across conditions is useZeros = FALSE in order to ensure the nonzero counts are scaled appropriately. The reason for this is demonstrated in Fig. 3 for a single-gene example in the H1-1 M and H1-4 M datasets.

In this example, since the two conditions contain the exact same cells, the normalized counts should have the same mean across the two groups. When useZeros=FALSE, SCnorm estimates across condition scale factors based on the nonzero expression counts and the resulting nonzero means are equal (Fig. 3a). This will lead to an appropriate call of *not* differentially expressed by methods such as MAST [12] and scDD [13]. Figure 3b demonstrates that if the DE method includes zeros in the differential expression test, then the overall mean of the two groups will not be equal. This effect is more pronounced in this example since the two conditions have very different proportions of zeros. To avoid an incorrect call of differential expression, the argument

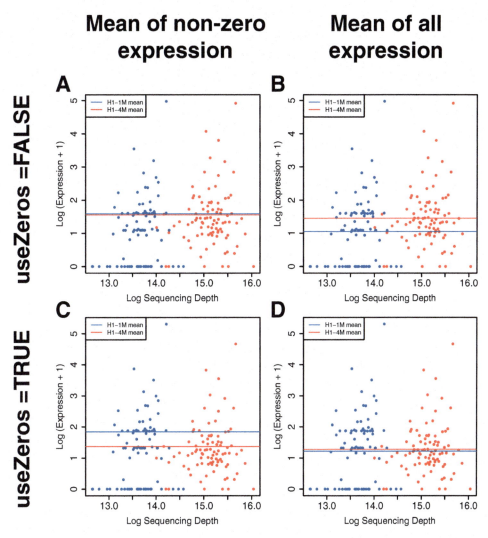

**Fig. 3** For a single gene, the log of the normalized counts for both the H1-1 M (blue) and H1-4 M (red) datasets versus log sequencing depth are shown. A constant of one was added to the counts before taking the log to highlight the zero counts. The top and bottom rows contain the normalized counts when useZeros=FALSE or useZeros=TRUE, respectively. The left column shows the condition-specific means calculated on the nonzero counts only. The right column shows the condition-specific means calculated over all counts

useZeros=TRUE should be used instead and SCnorm will scale across conditions including the zeros in the scale factor estimation. Under this option, the means of the nonzero counts will not be equal if the proportion of zeros is different across conditions and will result in the nonzero mean of the group with more zeros to appear higher (Fig. 3c), however the gene means including zeros will be equal (Fig. 3d).

The results of the across condition scaling under the two options are most critical when the conditions have very different proportions of zeros. The useZeros=FALSE option is most optimal for DE methods which treat zeros separately, while useZeros=TRUE is most optimal for DE methods that consider all counts together.

4. If spike-ins are present in the data, they may be used to perform the across condition scaling. First, we want to check that the spike-ins do not take up an extreme proportion of total counts. Following with the data above, we can check this directly:

```
> spikes <- grep("ERCC-", rownames(allData), value=TRUE)
> spikeRatio <- colSums(allData[spikes,]) / colSums(allData)
> head(sort(spikeRatio, decreasing = TRUE))
P96_H1b5s_026 P24_H1b5s_026 P24_H1b5s_061 P96_H1b5s_061
P24_H1b5s_049
    0.1975064     0.1972390     0.1703899     0.1677037
0.1458059
P96_H1b5s_049
    0.1433341
```

Cells with more than 20% of spike-ins are typically removed prior to analysis during quality control. In addition, it should be checked that the proportion of spike-ins is not drastically different across cells or groups. Here we check the average proportion across the two groups:

```
> mean(spikeRatio[colnames(h1cells.1M)])
0.04553682
> mean(spikeRatio[colnames(h1cells.4M)])
0.0460152
```

Specifically, for SCnorm, we want to make sure the spike-ins span the range of expression and are representative of biological genes. We check this by calculating how many spike-ins are in each expression group:

```
> spikeGroups <- subset(cdr.1M[[1]], Gene %in% spikes)
> table(spikeGroups$Group)
 1 2 3 4 5 6 7 8 9 10
 5 3 6 4 7 4 4 3 3 16
> spikeGroups <- subset(cdr.4M[[1]], Gene %in% spikes)
> table(spikeGroups$Group)
 1 2 3 4 5 6 7 8 9 10
 7 3 3 5 6 6 4 5 3 16
```

Not surprisingly, most of the spike-ins are in the tenth expression group which is the most highly expressed genes. In this experiment, they appear to span the range of expression well and may be considered for across condition scaling.

```
> myNormData.spikesUsed <- SCnorm(Data = allData, Conditions =
         Conditions, PrintProgressPlots = TRUE, useSpikes=TRUE)
```

The use of spike-ins for normalization should be carefully considered in practice; they must be of high-quality across cells and representative of biological genes [8, 15].

## References

1. Robinson MD, Oshlack A (2010) A scaling normalization method for differential expression analysis of RNA-seq data. Genome Biol 11 (3):R25

2. Bacher R et al (2017) SCnorm: robust normalization of single-cell RNA-seq data. Nat Methods 14(6):584–586

3. Anders S, Huber W (2010) Differential expression analysis for sequence count data. Genome Biol 11(10):R106

4. Li B, Dewey CN (2011) RSEM: accurate transcript quantification from RNA-Seq data with or without a reference genome. BMC Bioinformatics 12:323

5. Risso D, Ngai J, Speed TP, Dudoit S (2014) Normalization of RNA-seq data using factor analysis of control genes or samples. Nat Biotechnol 32(9):896–902

6. Kaufman L, Rousseeuw P (1987) Clustering by means of medoids. In Statistical Data Analysis Based on the L1 Norm and Related Methods (pp. 405–416). North-Holland; Amsterdam.

7. Mächler M, Rousseeuw P, Struyf A, Hubert M, Hornik K (2012) Cluster: cluster analysis basics and extensions. R package version, 1(2), 56

8. Bacher R, Kendziorski C (2016) Design and computational analysis of single-cell RNA-sequencing experiments. Genome Biol 17(1):63

9. Morgan M, Obenchain V, Hester J, Pagès H (2017) SummarizedExperiment: summarizedExperiment container. https://bioconductor.org/packages/release/bioc/html/SummarizedExperiment.html

10. Leng N et al (2013) EBSeq: an empirical Bayes hierarchical model for inference in RNA-seq experiments. Bioinformatics 29(8):1035–1043

11. Love MI, Huber W, Anders S (2014) Moderated estimation of fold change and dispersion for RNA-seq data with DESeq2. Genome Biol 15(12):550

12. Finak G et al (2015) MAST: a flexible statistical framework for assessing transcriptional changes and characterizing heterogeneity in single-cell RNA sequencing data. Genome Biol 16 (1):278

13. Korthauer KD et al (2016) A statistical approach for identifying differential distributions in single-cell RNA-seq experiments. Genome Biol 17(1):222

14. Risso D, Schwartz K, Sherlock G, Dudoit S (2011) GC-content normalization for RNA-seq data. BMC Bioinformatics 12(1):480

15. Stegle O, Teichmann SA, Marioni JC (2015) Computational and analytical challenges in single-cell transcriptomics. Nat Rev Genet 16 (3):133–145

# Chapter 3

# Analysis of Technical and Biological Variability in Single-Cell RNA Sequencing

## Beomseok Kim, Eunmin Lee, and Jong Kyoung Kim

## Abstract

Profiling the transcriptomes of individual cells with single-cell RNA sequencing (scRNA-seq) has been widely applied to provide a detailed molecular characterization of cellular heterogeneity within a population of cells. Despite recent technological advances of scRNA-seq, technical variability of gene expression in scRNA-seq is still much higher than that in bulk RNA-seq. Accounting for technical variability is therefore a prerequisite for correctly analyzing single-cell data. This chapter describes a computational pipeline for detecting highly variable genes exhibiting higher cell-to-cell variability than expected by technical noise. The basic pipeline using the scater and scran R/Bioconductor packages includes deconvolution-based normalization, fitting the mean-variance trend, testing for nonzero biological variability, and visualization with highly variable genes. An outline of the underlying theory of detecting highly variable genes is also presented. We illustrate how the pipeline works by using two case studies, one from mouse embryonic stem cells with external RNA spike-ins, and the other from mouse dentate gyrus cells without spike-ins.

**Key words** Single-cell RNA-seq, Technical variability, Biological variability, Cell-to-cell variability, Gene expression noise, Highly variable genes

## 1 Introduction

Since the first paper showing the feasibility of characterizing the transcriptomes of individual cells was published by Tang and colleagues in 2009 [1], single-cell RNA sequencing (scRNA-seq) has been widely applied to dissect cellular heterogeneity within a population of cells [2]. The early experimental approaches for scRNA-seq were limited in terms of both technical noise and throughput of cells [3]. Two technological breakthroughs have resolved these two issues. First, amplifying a minute amount of mRNA in a single cell by either PCR or in vitro transcription leads to a high level of technical cell-to-cell variability in gene expression [4]. A molecular barcoding approach originally developed for bulk RNA-seq [5],

Beomseok Kim and Eunmin Lee contributed equally to this work.

Guo-Cheng Yuan (ed.), *Computational Methods for Single-Cell Data Analysis*, Methods in Molecular Biology, vol. 1935, https://doi.org/10.1007/978-1-4939-9057-3_3, © Springer Science+Business Media, LLC, part of Springer Nature 2019

which barcodes each mRNA molecule with randomly synthesized oligonucleotides (also known as unique molecular identifier (UMI)), has been adapted to protocols for scRNA-seq to reduce technical noise arising from amplification bias [6]. Second, single-cell isolation methods combining microfluidics technologies with combinatorial cellular indexing have enabled the transcriptome of tens of thousands of single cells to be profiled [7, 8].

Despite recent technological advances of scRNA-seq, several technical challenges remain to be overcome. Both inefficient reverse transcription and insufficient sequencing depth per cell result in dropout events and sparsity in a gene expression matrix, which are the major cause of much higher technical variability of gene expression in scRNA-seq than in bulk RNA-seq [4]. Stochastic gene expression arising from random biochemical reactions and transcriptional bursting is also responsible for a higher level of biological variability in scRNA-seq [9]. Accounting for both technical and biological variability is therefore a prerequisite for correctly analyzing single-cell data.

To overcome this problem, a general workflow for analyzing scRNA-seq data has been widely adopted [10]. Given a gene expression matrix, we first remove poor-quality cells based on various quality control metrics [11]. After properly normalizing raw read or UMI counts by correcting for the effects of cell-level technical covariates, we identify highly variable genes showing higher cell-to-cell variability than expected by technical variability [4]. These genes are then used to visualize high-dimensional single-cell data in a two or three-dimensional space, and to cluster cells into groups sharing similar expression patterns. Finally, based on the identified cell clusters, we identify marker genes for each cluster and reconstruct developmental trajectories by ordering cells in pseudo-time [12].

In this chapter, we describe a pipeline for identifying highly variable genes from scRNA-seq data with or without external RNA spike-ins. We first provide an outline of the underlying theory for accounting for technical variability in terms of variance decomposition. We then describe how highly variable genes can be detected with the help of external RNA spike-ins. Since external spike-ins are not available in high-throughput scRNA-seq data, alternative methods inferring technical noise from the overall mean-variance trend of endogenous genes are also discussed.

## 2 Materials

Our protocol for analyzing technical and biological variability from scRNA-seq data requires the R programming language for statistical computing and optionally an integrated development environment for R (e.g., RStudio).

## 2.1 R/Bioconductor Packages

The following R/Bioconductor packages are required:

1. scater [13]: Single-cell analysis toolkit for gene expression data in R (https://bioconductor.org/packages/release/bioc/html/scater.html),

2. scran [14]: Methods for single-cell RNA-seq data analysis (https://bioconductor.org/packages/release/bioc/html/scran.html).

The Bioconductor packages can be installed by using biocLite():

```
source("https://bioconductor.org/biocLite.R")
biocLite("scater")
biocLite("scran")
```

## 2.2 scRNA-seq Datasets

To demonstrate how the pipeline works with or without the help of spike-ins, we use two different scRNA-seq datasets: (1) mouse embryonic stem cells (mESCs) with external RNA spike-ins [15], and (2) micro-dissected cells from mouse dentate gyrus without external RNA spike-ins [16].

### 2.2.1 scRNA-seq Dataset with Spike-Ins

This dataset consists of 704 mESCs cultured in three different conditions: serum + LIF (three replicates), 2i + LIF (four replicates) and alternative 2i + LIF (two replicates). All the mESCs passed cell-level quality control criteria. For each replicate, 96 single cells were captured with the Fluidigm C1 system and 92 ERCC RNA spike-ins were added to cell lysate. The cDNA and Illumina library were prepared with the SMARTer Kit and the Nextera XT Kit, respectively. Of four replicates of 2i-cultured mESCs, we will use two replicates: 2i2 and 2i3. The first replicate (2i2) has poor-quality spike-ins while the other replicate (2i3) has good-quality spike-ins. The raw read count table is publicly available at https://www.ebi.ac.uk/teichmann-srv/espresso.

### 2.2.2 scRNA-seq Dataset without Spike-Ins

This dataset contains 5454 cells from mouse developing dentate gyrus, which were sampled at four postnatal time points (P12, P16, P24, and P35). Cells were dissociated and captured with the 10× Genomics Chromium platform on two experimental days: day1 (P12 and P35), and day2 (P16 and P24). Low-quality cells and doublets were filtered out. The UMI count table and the corresponding annotation data are publicly available from the Gene Expression Omnibus (GEO) at the accession number of GSE95315.

## 3 Methods

Accounting for technical variability in scRNA-seq data has become an essential component of statistical methods for identifying highly variable genes and quantifying biological variability. Most such methods decompose the total variance into technical and biological components based on the law of total variance [4, 14, 17]. We provide a brief introduction to the common statistical framework for these methods.

### 3.1 A Statistical Framework to Account for Technical Variability

Suppose that $x_{ij}$ is a random variable denoting the unknown number of transcripts (or concentration) of gene $i$ in cell $j$. The number of transcripts (or concentration) of gene $i$ in cell $j$ available for sequencing after cell lysis, reverse transcription, and cDNA amplification steps is denoted by $z_{ij}$. We also denote by $k_{ij}$ as the observed read or UMI count of gene $i$ in cell $j$. By the general theorem of variance decomposition [18], the variance of $k_{ij}$ can be decomposed into:

$$\text{Var}\left[k_{ij}\right] = E\left[\text{Var}\left[k_{ij}|z_{ij}, x_{ij}\right]\right] + E\left[\text{Var}\left[E\left[k_{ij}|z_{ij}, x_{ij}\right]|x_{ij}\right]\right] + \text{Var}\left[E\left[k_{ij}|x_{ij}\right]\right].$$

The first term explains the technical variability arising from sequencing noise, which is usually modeled using a Poisson process [19]. The second term quantifies the technical variability generated by stochastic mRNA loss during the single-cell library preparation steps, which is a major source of technical variability. The last term quantifies the biological variability. The basic idea of identifying highly variable genes is to find genes whose observed variance is not dominated by the first two technical variability terms. In other words, highly variable genes can be defined as genes showing significant nonzero biological variance.

In principle, the technical variability terms should be estimated from external RNA spike-ins since we can eliminate the biological cell-to-cell variability of $x_{ij}$ from the decomposition formula. Then, the variance of $k_{ij}$ for spike-ins can be simplified by the law of total variance:

$$\text{Var}\left[k_{ij}\right] = E\left[\text{Var}\left[k_{ij}|z_{ij}\right]\right] + \text{Var}\left[E\left[k_{ij}|z_{ij}\right]\right].$$

It should be noted that the above two terms correspond to the first two technical variability terms if we assume $x_{ij}$ is a fixed and known quantity, which is a reasonable assumption for spike-ins (*see* **Note 1**). To plug-in the estimated technical variability of spike-ins into that of endogenous genes, we make an assumption that the technical variance of spike-ins is a nonlinear function of their mean expression levels. By fitting a curve to the mean-variance (or variance derived quantities like coefficient of variation) data of spike-ins using a nonlinear regression function, we can estimate the average technical variance of each endogenous gene at the given mean expression level. The biological variance of endogenous genes can be estimated by subtracting the average technical variance from the total observed variance. If spike-ins have poor quality or are not available, we directly estimate the mean-variance trend of technical noise from endogenous genes by assuming that most genes are dominated by technical variability.

To test whether the biological variance of $k_{ij}$ is equal to a specified value (usually set to 0), the ratio of the sample variance of the normalized counts to the expected variance under the null hypothesis is usually used as a test statistic. The test statistic under the null hypothesis follows a chi-squared distribution if we assume that the normalized counts follow a normal distribution. To make this normality assumption more reasonable, some methods (e.g., scran) take log-transformed normalized counts as their input expression values.

### 3.2 Identifying Highly Variable Genes with External RNA Spike-Ins

#### 3.2.1 Data Loading and Normalization

We first load all R and Bioconductor packages we need in this protocol, and then load the raw read count table for mESCs from counttable_es.txt.

```
library(scater)
library(scran)
ct <- as.matrix(read.table("counttable_es.txt",
                    sep = " ",
                    header = T,
                    row.names = 1,
                    check.names = FALSE))
```

From the count matrix, we select rows whose names start with "ENSMUSG" (Ensembl mouse gene ID) or "ERCC" (ERCC spike-in ID), and columns corresponding to one replicate of the 2i condition (2i3). The chosen count matrix is used to create a SingleCellExperiment object from scater which will serve as a data container compatible with many other Bioconductor packages including scran. The rows corresponding to spike-ins can be set using the isSpike function from scran.

```
ct <- ct[grepl("^ENSMUSG|^ERCC-", rownames(ct)), grepl("_2i_3",

colnames(ct))]

sceset <- SingleCellExperiment(assays = list(counts = ct))

isSpike(sceset, "ERCC") <- grepl("^ERCC-", rownames(sce))

sceset

## class: SingleCellExperiment

## dim: 38653 59

## metadata(0):

## assays(1): counts

## rownames(38653): ENSMUSG00000000001 ENSMUSG00000000003 ...

##    ERCC-00170 ERCC-00171

## rowData names(0):

## colnames(59): ola_mES_2i_3_10.counts ola_mES_2i_3_11.counts ...

##    ola_mES_2i_3_92.counts ola_mES_2i_3_96.counts

## colData names(0):

## reducedDimNames(0):

## spikeNames(0):
```

The raw read count of a gene in a cell is affected by cell-specific technical and biological factors. The cell-to-cell differences in the efficiency of cell lysis, capture efficiency of mRNA transcripts, and sequencing depth are known to affect the raw read counts. The total RNA content and cell size can also differ from cell to cell, which are biological factors affecting the amount of biological RNA available for sequencing and the raw read count. To eliminate the effects of these technical and biological factors, we normalize the raw read count by dividing each count by its cell-level size factor. Two different size factors are calculated with the computeSum-Factors and computeSpikeFactors functions from scran (*see* **Note 2**): biological size factors using endogenous genes and technical size factors using spike-ins. The biological size factors are computed by the deconvolution approach to deal with the sparsity of single-cell data [20]. Based on the assumption that a majority of genes are not differentially expressed, cell-specific biases arising from both technical and biological factors are normalized with the biological size factors. In contrast, the technical size factors are computed based on the spike-in counts to adjust for the effects of technical factors. Since the same amount of spike-ins are added to each cell, the cell-to-cell differences in total RNA content are not normalized with the technical size factors (*see* **Note 3**). Using the `normalize` function from scater, we normalize the raw read counts of endogenous genes with the biological size factors, and that of spike-ins with the technical size factors. The function calculates log2-transformed normalized expression values by adding a pseudo-count of 1, which are stored in `logcounts` or `exprs` of the returned `SingleCellExperiment` object.

```
sceset <- computeSumFactors(sceset, sizes = seq(10, 50, 5))

sceset <- computeSpikeFactors(sceset, general.use = FALSE)

sceset <- normalize(sceset)
```

*3.2.2 Detecting Highly Variable Genes*

From the log2-transformed normalized expression values, we first fit a curve to the mean-variance values of spike-ins with the `trendVar` function of scran (*see* **Note 4**).

```
var.fit <- trendVar(sceset, parametric = TRUE, use.spikes = TRUE)
```

The fitted value of the curve at a given mean expression level is used as an estimate of the technical component of the total variance of an endogenous gene expressed at the same mean level. The biological variance can be obtained by subtracting the estimated technical variance from the total observed variance for each endogenous gene. Under the null hypothesis that the total variance is equal to the estimated technical variance (or equivalently the biological variance is equal to 0), we can test whether the total variance is larger than the expected technical variance with the `decomposeVar` function of scran. Highly variable genes can be defined as genes showing significant nonzero biological variance. In this example, we consider a gene to be highly variable if it has a biological variance greater than 0.5 and its FDR is less than or equal to 0.05 (*see* **Note 5**).

```
var.out <- decomposeVar(sceset, var.fit)
hvg <- var.out[which(var.out$FDR <= 0.05 & var.out$bio > .5 ),]
nrow(hvg)

## [1] 2109

head(hvg)

## DataFrame with 6 rows and 6 columns
##                                    mean                total
bio
##                               <numeric>            <numeric>            <numer
ic>
## ENSMUSG00000000001 6.87377936714626 7.63664354974175


4.94614195802085

## ENSMUSG00000000028 7.16813281638306 8.36720349968789

6.19082343482952

## ENSMUSG00000000131 8.42914208049494 3.41985542516078

2.60116398672207

## ENSMUSG00000000171 8.87511548573431 2.87241703701955

2.2966896068407

## ENSMUSG00000000278 9.14542287426345  2.1818210345152

1.71403288145084

## ENSMUSG00000000295 6.33472585927806 6.86610866121096

2.89637950173032

##                                    tech              p.value
##                               <numeric>            <numeric>
## ENSMUSG00000000001    2.6905015917209 3.78117460395952e-12
```

```
## ENSMUSG00000000028  2.17638006485837 3.48532179254256e-21

## ENSMUSG00000000131 0.818691438438704 2.24109579972162e-24

## ENSMUSG00000000171 0.575727430178854 1.83553665593285e-32

## ENSMUSG00000000278 0.467788153064363 3.51389775367481e-29

## ENSMUSG00000000295  3.96972915948064 0.000474348732430948

##                                      FDR

##                                  <numeric>

## ENSMUSG00000000001 1.55277820983262e-10

## ENSMUSG00000000028 2.65607694945126e-19

## ENSMUSG00000000131 2.01098169057812e-22

## ENSMUSG00000000171 2.67094826375195e-30

## ENSMUSG00000000278 4.18208059504488e-27

## ENSMUSG00000000295  0.00884495235554632
```

Figure 1a shows the mean-variance plot of the log2-transformed expression values of both endogenous genes (black points) and spike-ins (green points) in 2i3, where the blue line represents the fitted mean-variance trend based on the spike-ins and red points correspond to highly variable genes.

```
plot(y = var.out$total, x = var.out$mean,
     ylab = "Variance of log-expression", xlab = "Mean log-expression",
     pch = 16,cex = 0.3)
o <- order(var.out$mean)
lines(y = var.out$tech[o], x = var.out$mean[o],
      col = "dodgerblue", lwd = 2)
points(var.fit$mean, var.fit$var,
       col = "green", pch = 16, cex = 0.3)

points(y = var.out$total[var.out$FDR <= 0.05 & var.out$bio >= 0.5 ],
       x = var.out$mean[var.out$FDR <= 0.05 & var.out$bio >= 0.5 ],
       col = "red", pch = 16, cex = 0.3)
```

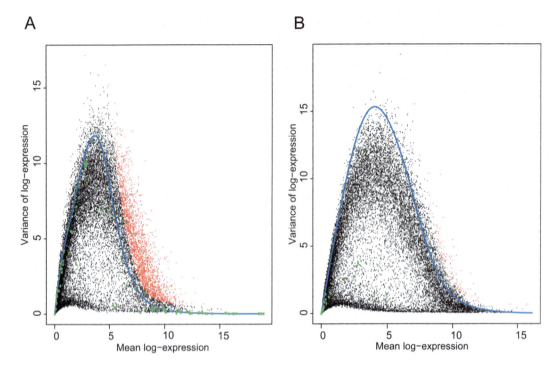

**Fig. 1** Mean-variance plots of log2-transformed normalized expression values of endogenous gens (black points) for 2i3 with a good quality of spike-ins (**a**) and 2i2 with a poor quality of spike-ins (**b**). Each green point represents a spike-in, and the blue line corresponds to the fitted mean-variance trend based on the spike-ins (**a**) or endogenous genes (**b**). Detected highly variable genes are colored by red

If we fail to add the same amount of spike-ins to each cell or the proportion of reads mapped to spike-ins is too low (<0.01% in 2i2 compared to 44% in 2i3), we can use an alternative approach by directly fitting the mean-variance trend to the normalized expression values of endogenous genes under the assumption that the technical components dominate the total variance in most genes (Fig. 1b, *see* **Note 6**):

```
var.fit <- trendVar(sceset, parametric = TRUE, use.spikes = FALSE)
```

### 3.3 Identifying Highly Variable Genes without External RNA Spike-Ins

In this section, we demonstrate how the basic pipeline described in Subheading 3.2 can be modified to detect highly variable genes for a large-scale scRNA-seq dataset without spike-ins. We first load the UMI count table for mouse dentate gyrus cells from `GSE95315_10X_expression_data.tab`, which is publicly available at ftp://ftp.ncbi.nlm.nih.gov/geo/series/GSE95nnn/GSE95315/suppl/GSE95315_10X_expression_data.tab.gz. The cell metadata containing the batch (experimental days) and cell type for each cell is loaded from `GSE95315_series_matrix`.

txt    available    at    ftp://ftp.ncbi.nlm.nih.gov/geo/series/
GSE95nnn/GSE95315/matrix/GSE95315_series_matrix.txt.gz.

```r
library(scater)
library(scran)
library(ggplot2)
ct <- as.matrix(read.table("GSE95315_10X_expression_data.tab",
                           sep = "\t",
                           header = T,
                           row.names = 1,
                           check.names = FALSE))
strs <- readLines("GSE95315_series_matrix.txt.gz")
dat <- read.csv(text=strs, sep = "\t",
                header = T,
                check.names=FALSE,
                skip = 30,
                nrows=length(strs) - 3 )
```

The count table and metadata is converted into a SingleCellExperiment object of scater.

```r
Annotation <- data.frame(celltype = t(dat)[2:ncol(dat), 13])
Annotation$celltype <- gsub("Neuroblast2", "Neuroblast_two", Annotati
on$celltype)
Annotation$celltype <- gsub("cell cluster: |1|2|glia-like| ", "", Ann
otation$celltype)
length(table( Annotation$celltype))

## [1] 22

cell.info <- data.frame(
        cell = colnames(ct),
        batch = sapply(colnames(ct), function(x) substr(x, 4, 7)),
        celltype = annotation[colnames(ct),"celltype"])
sceset <- SingleCellExperiment(
        assays = list(counts = ct), colData = cell.info)
sceset
```

```
## class: SingleCellExperiment

## dim: 14545 5454

## metadata(0):

## assays(1): counts

## rownames(14545): 0610007P14Rik 0610009B22Rik ... mt-Nd5 mt-Nd6

## rowData names(0):

## colnames(5454): 10X46_1_GCCTACACGGGAGT-1 10X46_1_AAGCACTGATGGTC-1

##    ... 10X43_1_ATGAAGGAATGCCA-1 10X46_1_GGACAGGATAGCGT-1

## colData names(3): cell batch celltype

## reducedDimNames(0):

## spikeNames(0):
```

We normalize the UMI counts with the deconvolution method implemented in the `computeSumFactors` function of scran. This method is based on the strong assumption that most genes are not differentially expressed across cells. However, this assumption is not valid for a large single-cell data composed of a mixture of different cell types. To weaken the assumption, cells are grouped into clusters based on their gene expression profiles with the `quickCluster` function of scran. For each cluster, the deconvolution-based normalization method is applied to compute the cluster-specific biological size factors, which are then rescaled across clusters.

```
clusters <- quickCluster(sceset)
sceset <- computeSumFactors(sceset, cluster = clusters)
sceset <- normalize(sceset)
```

As spike-ins are not present in this dataset, we make an assumption that the contribution of biological components to the total variance is marginal in most genes. Based on this assumption, we fit the mean-variance trend to the variances of log2-transformed normalized expression values of endogenous genes by setting

```
var.fit <- trendVar(sceset, method = "spline", parametric = TRUE,
                    use.spikes = FALSE, span = 0.2)
```

.

After we test whether the estimated biological variance is equal to 0, we define highly variable genes as ones with FDR ≤ 0.05 and biological variance ≥0.1.

```
var.out <- decomposeVar(sceset, var.fit)
hvg <- var.out[which(var.out$FDR <= 0.05 & var.out$bio >= 0.1),]

nrow(hvg)

## [1] 312

head(hvg)

## DataFrame with 6 rows and 6 columns

##                         mean                total                bio

##                    <numeric>            <numeric>          <numeric>

## 1810037I17Rik 0.680310681114725 0.908861494285629

0.22438946719247

## 2010300C02Rik 0.602083745312206 0.785529562526665

0.158613308891032

## 2900055J20Rik 0.642681674116672   0.7894952004639

0.132051146817701

## 6330403K07Rik 0.760365192522513 0.873966446743902

0.136031205136787
```

```
## Abr           0.723940378793588 0.852462429491801 0.138174667075084

## Acsbg1         0.15950685385264 0.404043372590482 0.205280822499748

##                         tech           p.value                FDR

##                    <numeric>         <numeric>          <numeric>

## 1810037I17Rik 0.68447202709316 8.49384128900126e-55

1.2317340134449e-53

## 2010300C02Rik 0.626916253635633 9.13838303865434e-35

9.65972247799617e-34

## 2900055J20Rik 0.657444053646199  2.9741904700187e-23

2.45235829855e-22

## 6330403K07Rik 0.737935241607115 4.65272653177871e-20

3.49014478621564e-19

## Abr           0.714287762416717 8.60152901400342e-22

6.80681390145156e-21

## Acsbg1         0.198762550090734                  0                  0
```

In Fig. 2, we plot the mean-variance relationship of endogenous genes (black points), where red points represent highly variable genes, and the blue line corresponds to the fitted trend.

```
plot(y = var.out$total, x = var.out$mean, pch = 16, cex = 0.3,
    ylab = "Variance of log-expression", xlab = "Mean log-expression")
o <- order(var.out$mean)
lines(y = var.out$tech[o], x = var.out$mean[o],
    col = "dodgerblue", lwd = 2)
points(y = var.out$total[var.out$FDR <= 0.05 & var.out$bio >= 0.1],
    x = var.out$mean[var.out$FDR <= 0.05 & var.out$bio >= 0.1],
    col="red", pch = 16, cex = 0.3)
```

Highly variable genes are informative as they vary across different cell types or subpopulations. Thus, they can be used as an input set of feature genes to visualize the high-dimensional single-cell

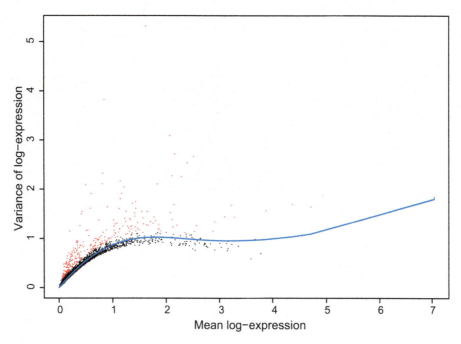

**Fig. 2** A mean variance plot of log2-transformed normalized expression values of genes in dentate gyrus cells. The blue line represents the fitted mean-variance trend, and highly variable genes are marked in red

data in a two- or three-dimensional space, or to cluster cells into groups showing similar gene expression profiles. We illustrate that highly variabe genes are useful for visualizing subpopulation structure associated with the known cell types, using t-distributed stochastic neighbor embedding (t-SNE) [21] implemented in the runTSNE function of scater. As a control, we also generate the t-SNE plot using all genes.

```
sce.tsne <- runTSNE(sceset, feature_set = rownames(sceset), rand_seed
= 123456)

sce.tsne.hvg <- runTSNE(sceset, feature_set = rownames(hvg), rand_see
d = 123456)
```

Figure 3 shows the t-SNE plots using all genes (Fig. 3a, c) or highly variable genes (Fig. 3b, d, *see* **Note 7**). Cell types inferred from [16] are overlaid in Fig. 3a, b. Two experimental days ("43_1" and "46_1") are also overlaid in Fig. 3c, d to look for batch effects caused by experimental days.

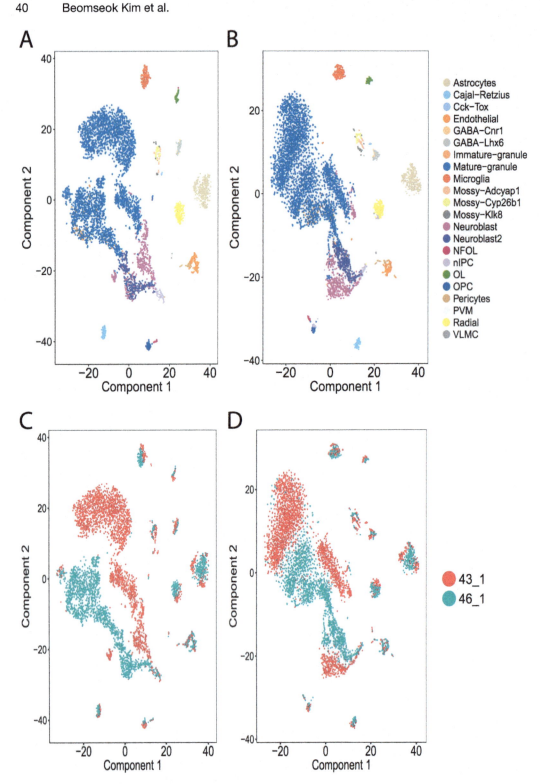

**Fig. 3** t-SNE plots of dentate gyrus cells using all genes (**a**, **c**) or highly variable genes (**b**, **d**). Cells are clustered by the annotated cell types (**a**, **b**), and partially by the experimental days (**c**, **d**)

```r
df <- data.frame(x = sce.tsne@reducedDims$TSNE[, 1],
                 y = sce.tsne@reducedDims$TSNE[, 2],
                 expression = sce.tsne@colData$celltype)
colors <- c("#E5D8BD", "#A6CEE3", "#B3CDE3", "#FDB462", "#FBD9A6",
"#CCCCCC", "#FDC086", "#377EB8", "#FC8D62", "#FDCDAC", "#FFFF99",
"#999999", "#E78AC3", "#7570B3", "#E7298A", "#DECBE4", "#66A61E",
"#386CB0", "#E5C494", "#B2B2B2", "#FFED6F", "#B3B3B3")
ggplot() +
  geom_point(data = df, aes(x = x, y = y, colour = expression),
             size = 0.5) +
  scale_color_manual(values = colors)+
  ylab("Component 2") +
  xlab("Component 1") +
  guides(col = guide_legend(nrow = 22), override.aes = list(size = 5))
```

When we select all genes as an input set of feature genes, granule and neuroblast cells are largely separated by their experimental days. In contrast, when highly variable genes are selected, such separation is reduced, suggesting that highly variable genes are more useful for identifying biologically meaningful structure of high-dimensional scRNA-seq data than all expressed genes.

## 4  Notes

1. The external RNA spike-ins cannot be used to estimate the technical variability of stochastic RNA losses arising from inefficient cell lysis since the spike-ins are added to cell lysates.

2. One can change the number of cells per pool by setting the sizes argument of the computeSumFactors function. By default, the number of cells per pool is set to a range from 20 to 100. However, an error will occur when the total number of cells in the sample is less than this range. In this case, the maximum pool size should be smaller than the total number of cells as shown in Subheading 3.2. The recommended minimum pool size for sparse UMI data is 20, but smaller pool size may be possible for scRNA-seq data with high sequencing depth. When negative size factors are obtained for some cells, the range of pool size should be increased. Another approach is to filter out both lowly expressed genes and poor-quality cells with low total counts.

3. To preserve the cell-to-cell differences in total RNA content, we can set `general.use = TRUE` of the `computeSpikeFactors` function. In this case, both endogenous genes and spike-ins are normalized with the technical size factors estimated from the spike-ins.

4. If the dataset consists of multiple batches, the batch information should be incorporated into a design matrix, which can be passed into the `trendVar` function (e.g., `design = model.matrix(~replicates)`).

5. A similar approach based on the squared coefficient of variation (CV2) of [4] is also implemented in the `technicalCV2` function of scran.

6. An alternative approach, which is implemented in the `DM` function of scran, calculates the distance-to-median (DM) values after fitting a running median curve to the log-transformed CV2 of genes against to the log-transformed mean. For each gene, the DM value is defined as the distance between the observed CV2 and the fitted trend. Genes with high DM values can be identified as highly variable genes.

7. The t-SNE plots can be also generated with the `plotTSNE` function of scater. The code for t-SNE plots in the Subheading 3.3 is written for a better visualization.

## Acknowledgments

This work was supported by the National Research Foundation of Korea funded by the Ministry of Science, ICT and Future Planning (2017R1C1B2007843, 2017M3C7A1048448, 2017M3A9B6073099, 2017M3A9D5A01052447), and by Business for Cooperative R&D between Industry, Academy, and Research Institute funded by the Ministry of SMEs and Startups (C0452791).

## References

1. Tang F, Barbacioru C, Wang Y, Nordman E, Lee C, Xu N, Wang X, Bodeau J, Tuch BB, Siddiqui A, Lao K, Surani MA (2009) mRNA-Seq whole-transcriptome analysis of a single cell. Nat Methods 6(5):377–382. https://doi.org/10.1038/nmeth.1315

2. Tanay A, Regev A (2017) Scaling single-cell genomics from phenomenology to mechanism. Nature 541(7637):331–338. https://doi.org/10.1038/nature21350

3. Kolodziejczyk AA, Kim JK, Svensson V, Marioni JC, Teichmann SA (2015) The technology and biology of single-cell RNA sequencing. Mol Cell 58(4):610–620. https://doi.org/10.1016/j.molcel.2015.04.005

4. Brennecke P, Anders S, Kim JK, Kolodziejczyk AA, Zhang X, Proserpio V, Baying B, Benes V, Teichmann SA, Marioni JC, Heisler MG (2013) Accounting for technical noise in single-cell RNA-seq experiments. Nat Methods 10(11):1093–1095. https://doi.org/10.1038/nmeth.2645

5. Kivioja T, Vaharautio A, Karlsson K, Bonke M, Enge M, Linnarsson S, Taipale J (2011)

Counting absolute numbers of molecules using unique molecular identifiers. Nat Methods 9 (1):72–74. https://doi.org/10.1038/nmeth. 1778

6. Islam S, Zeisel A, Joost S, La Manno G, Zajac P, Kasper M, Lonnerberg P, Linnarsson S (2014) Quantitative single-cell RNA-seq with unique molecular identifiers. Nat Methods 11(2):163–166. https://doi.org/10. 1038/nmeth.2772

7. Macosko EZ, Basu A, Satija R, Nemesh J, Shekhar K, Goldman M, Tirosh I, Bialas AR, Kamitaki N, Martersteck EM, Trombetta JJ, Weitz DA, Sanes JR, Shalek AK, Regev A, McCarroll SA (2015) Highly parallel genome-wide expression profiling of individual cells using Nanoliter droplets. Cell 161 (5):1202–1214. https://doi.org/10.1016/j. cell.2015.05.002

8. Klein AM, Mazutis L, Akartuna I, Tallapragada N, Veres A, Li V, Peshkin L, Weitz DA, Kirschner MW (2015) Droplet barcoding for single-cell transcriptomics applied to embryonic stem cells. Cell 161 (5):1187–1201. https://doi.org/10.1016/j. cell.2015.04.044

9. Kim JK, Marioni JC (2013) Inferring the kinetics of stochastic gene expression from single-cell RNA-sequencing data. Genome Biol 14 (1):R7. https://doi.org/10.1186/gb-2013-14-1-r7

10. Stegle O, Teichmann SA, Marioni JC (2015) Computational and analytical challenges in single-cell transcriptomics. Nat Rev Genet 16 (3):133–145. https://doi.org/10.1038/ nrg3833

11. Ilicic T, Kim JK, Kolodziejczyk AA, Bagger FO, McCarthy DJ, Marioni JC, Teichmann SA (2016) Classification of low quality cells from single-cell RNA-seq data. Genome Biol 17:29. https://doi.org/10.1186/s13059-016-0888-1

12. Trapnell C, Cacchiarelli D, Grimsby J, Pokharel P, Li S, Morse M, Lennon NJ, Livak KJ, Mikkelsen TS, Rinn JL (2014) The dynamics and regulators of cell fate decisions are revealed by pseudotemporal ordering of single cells. Nat Biotechnol 32(4):381–386. https:// doi.org/10.1038/nbt.2859

13. McCarthy DJ, Campbell KR, Lun ATL, Wills QF (2017) Scater: pre-processing, quality control, normalization and visualization of single-cell RNA-seq data in R. Bioinformatics 33 (8):1179–1186. https://doi.org/10.1093/ bioinformatics/btw777

14. Lun ATL, McCarthy DJ, Marioni JC (2016) A step-by-step workflow for low-level analysis of single-cell RNA-seq data with bioconductor. F1000Res 5:2122. https://doi.org/10. 12688/f1000research.9501.2

15. Kolodziejczyk AA, Kim JK, Tsang JC, Ilicic T, Henriksson J, Natarajan KN, Tuck AC, Gao X, Buhler M, Liu P, Marioni JC, Teichmann SA (2015) Single cell RNA-sequencing of pluripotent states unlocks modular transcriptional variation. Cell Stem Cell 17(4):471–485. https:// doi.org/10.1016/j.stem.2015.09.011

16. Hochgerner H, Zeisel A, Lonnerberg P, Linnarsson S (2018) Conserved properties of dentate gyrus neurogenesis across postnatal development revealed by single-cell RNA sequencing. Nat Neurosci 21(2):290–299. https://doi.org/10.1038/s41593-017-0056-2

17. Kim JK, Kolodziejczyk AA, Ilicic T, Teichmann SA, Marioni JC (2015) Characterizing noise structure in single-cell RNA-seq distinguishes genuine from technical stochastic allelic expression. Nat Commun 6:8687. https://doi.org/ 10.1038/ncomms9687

18. Bowsher CG, Swain PS (2012) Identifying sources of variation and the flow of information in biochemical networks. Proc Natl Acad Sci U S A 109(20):E1320–E1328. https://doi.org/ 10.1073/pnas.1119407109

19. Marioni JC, Mason CE, Mane SM, Stephens M, Gilad Y (2008) RNA-seq: an assessment of technical reproducibility and comparison with gene expression arrays. Genome Res 18(9):1509–1517. https://doi. org/10.1101/gr.079558.108

20. Lun ATL, Bach K, Marioni JC (2016) Pooling across cells to normalize single-cell RNA sequencing data with many zero counts. Genome Biol 17:75. https://doi.org/10. 1186/s13059-016-0947-7

21. Van der Maaten L, Hinton GE (2008) Visualizing Data using t-SNE. J Mach Learn Res 9:2579–2605

# Chapter 4

## Identification of Cell Types from Single-Cell Transcriptomic Data

**Karthik Shekhar and Vilas Menon**

### Abstract

Unprecedented technological advances in single-cell RNA-sequencing (scRNA-seq) technology have now made it possible to profile genome-wide expression in single cells at low cost and high throughput. There is substantial ongoing effort to use scRNA-seq measurements to identify the "cell types" that form components of a complex tissue, akin to taxonomizing species in ecology. Cell type classification from scRNA-seq data involves the application of computational tools rooted in dimensionality reduction and clustering, and statistical analysis to identify molecular signatures that are unique to each type. As datasets continue to grow in size and complexity, computational challenges abound, requiring analytical methods to be scalable, flexible, and robust. Moreover, careful consideration needs to be paid to experimental biases and statistical challenges that are unique to these measurements to avoid artifacts. This chapter introduces these topics in the context of cell-type identification, and outlines an instructive step-by-step example bioinformatic pipeline for researchers entering this field.

**Key words** Single-cell RNA-sequencing, Transcriptomic classification, Cell-type identification, Cell taxonomy, Clustering, Unsupervised machine learning, Cross-species comparison of cell-types

---

## 1 Introduction

The human body contains approximately 40 trillion cells, which exhibit a breathtaking diversity of form and function [1]. Classifying these cells into "types" is increasingly viewed as a foundational requirement to gain a detailed understanding of how tissues function and interact, and to uncover specific mechanisms that underlie pathological states [2]. Provisionally, cells of a particular type share a common identity defined by multiple, measurable properties pertaining to tissue location, function, signaling properties, morphology, electrophysiological response, molecular composition, and physicochemical interaction with other cell types (see below). Knowledge of what cell types exist and what features distinguish them will (a) facilitate genetic access to specific types so they can be labeled and manipulated in model organisms and culture systems,

Guo-Cheng Yuan (ed.), *Computational Methods for Single-Cell Data Analysis*, Methods in Molecular Biology, vol. 1935, https://doi.org/10.1007/978-1-4939-9057-3_4, © Springer Science+Business Media, LLC, part of Springer Nature 2019

(b) provide a framework to investigate the staggering cellular heterogeneity that abounds in organisms, (c) provide mechanistic insight into the generation of this heterogeneity during early development, (d) provide a framework for rationally improving in vitro-derived cell types, (e) facilitate cross-species comparisons [3], and (f) implicate roles for specific cell types and their interactions [4] in complex diseases [5, 6].

Although the genomes of complex mammals contain ~30,000 genes (and their multiple isoforms), the expression patterns of these genes are not all independent of each other. Gene regulatory processes induce correlations between the expression levels of genes, which in turn results in a "modular" structure of the transcriptome [7]. One consequence of this modularity is that the molecular states of cells occupy a low dimensional subspace (often referred to as a "manifold") within the full space of gene expression. Advances in single-cell RNA-sequencing (scRNA-seq) technology have enabled cell types to be defined using the transcriptomic state of thousands of individual cells [8–10]. In addition, the development of single-nucleus profiling techniques has allowed for thorough investigations of frozen and banked tissue, including challenging tissues such as adult human brain sections [11, 12]. A flurry of recent work has shown that unbiased classification of single-cell transcriptomes using computational methods rooted in clustering and dimensionality reduction not only recovers classically defined subsets of cells, but also enables discovery of novel types with unknown functional roles [13–15]. Our goal is to introduce the reader to the conceptual [16] and computational [17] challenges of scRNA-seq data analysis, followed by a description of a basic practical workflow of scRNA-seq analysis using the R statistical language.

## 1.1 What Is a Cell Type?

While every cell is unique, experience of biologists over many years has suggested that cells can be organized into groups based on shared features that are quantifiable. This categorization makes possible systematic and reproducible analyses of complex tissues, similar to the concept of "species," which greatly simplifies the diversity of organisms into an interpretable taxonomy, while not denying the individuality of any single member [18]. Features used to define cell types include lineage, location, morphology, activity, interactions with other cell types, epigenetic state, responsiveness to certain signals, and molecular composition (including mRNA and protein levels) [16].

scRNA-seq-based cell classification involves partitioning the data into "clusters" of single cells, wherein each cluster is defined by a unique gene expression "signature" relative to other clusters, and therefore, represents a putative cell type. It must be noted, however, that a computationally defined cluster may not necessarily correspond 1:1 to a cell type, as the molecular state of the cell assayed by scRNA-seq may not necessarily reflect all of the features noted above. Moreover, certain molecular attributes are more transient than others during the lifetime of a cell, necessitating a distinction between a cell's type (its principal identity) and its current "state" (e.g., temporary changes in firing rate of neuron during "up" and "down" states, or different levels of secretory activity of endocrine cells). scRNA-seq may resolve different "states" of the same cell type if their transcriptional signatures are sufficiently distinct, and collapse two distinct, but closely related types if the molecules that specified their identities during early development are no longer expressed during the stage of the experiment.

Thus, even when restricted to the molecular state, the difference between cell "types" and "states" is not resolvable through RNA-seq alone, and may require examination in other modalities, such as those that capture information about the cell's epigenetic state or its dynamical responses. Taken together, these caveats warrant caution in the interpretation of scRNA-seq data, especially in the context of identifying cell types exclusively from transcriptomic information. The notion of cell types is being constantly refined through ongoing work as part of large-scale projects such as the Human Cell Atlas [2] and the BRAIN initiative [19].

## 1.2 A Brief Overview of scRNA-Seq

scRNA-seq is not a single method, but a suite of protocols, each with its strengths and limitations [20]. Currently, every scRNA-seq protocol consists of three steps (Fig. 1): (1) single-cell capture and barcoding, (2) library preparation, and (3) sequencing. Current protocols isolate single cells by tissue dissociation, followed by either fluorescence- activated cell sorting (FACS) into separate wells on a plate or capturing individual cells in microfluidic chambers, microwells, or individual droplets. Prior to single-cell capture, the dissociated cells can be optionally taken through a sorting step using FACS or magnetic activated cell-sorting (MACS) to enrich or deplete cells expressing a specific combination of markers. Library preparation involves reverse transcribing mRNA into cDNA and amplifying, either using polymerase chain reaction (PCR) or in vitro transcription (IVT). Recently developed protocols tag transcripts during the capture stage (step 1 above) with unique molecular identifiers (UMIs), which are random nucleotide sequences [21]. Every captured transcript is, in principle, tagged with a distinct UMI, which enables downstream correction of amplification biases. The amplified cDNA is then fragmented, followed by the addition of molecular adapters at the end of amplicon fragments that allow for high-throughput sequencing. Libraries can either retain the full length of every transcript or tag either the 3′ or the 5′ end of each mRNA—the choice is informed by further considerations. Sequencing is generally highly multiplexed, and can either be single-end or paired-end depending on upstream choices. An important consideration can be the depth of sequencing per cell, which is often related to the number of cells profiled [22].

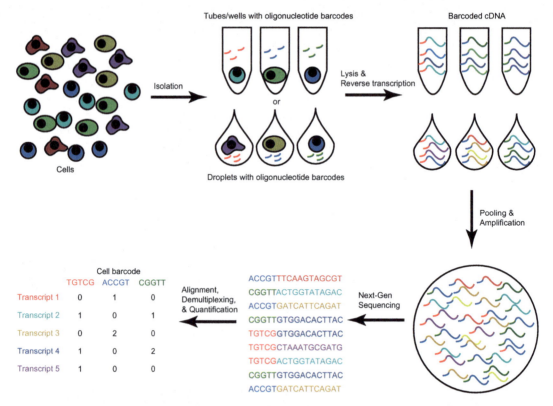

**Fig. 1** General experimental workflow for single-cell RNA-sequencing, as described in detail in the text, starting from cell isolation and extending through to the generation of counts tables showing the detection of each gene in each cell

***1.3 Batch Effects in scRNA-Seq Analysis***

Data-driven identification of cell types can be confounded by batch effects, which result from minor, but systematic differences between experimental replicates prepared either at different times, using different reagent batches, different experimenters, or a combination of the three [23]. Batch effects can result in variation in the transcriptomic state of identical cell types across different replicates due to technical factors; when such effects are strong, cells can cluster by batch identity rather than biological identity. Batch effects can also arise if in addition to transcriptional differences, the frequencies of specific cell types are different across batches [24, 25]. If different biological conditions of interest (e.g., control vs. perturbation) or different sample sources (e.g., biopsies from cancer patients) are processed in different batches, it is statistically impossible to deconvolve biological versus technical effects. While batch effects can be mitigated through careful experimental design involving an even distribution of different biological conditions across experimental batches ("block design"), although this may not always be logistically feasible if delays in sample processing can compromise quality. In such circumstances, cell-types and molecular signals identified in a single experimental batch must be treated with suspicion and results should only be believed if they are supported across multiple independent replicates, or in other data modalities. Detecting and correcting batch effects is an ongoing area of computational innovation, and a number of approaches have been recently proposed [24–26].

Future promising avenues of research involve the integration of scRNA-seq data directly with other data modalities. In particular, recent developments linking RNA-seq to spatial location (such as FISSEQ [27] and "Spatial Transcriptomics" [28]), combined with the advent of high-resolution and expansion microscopy, are on the verge of collecting transcriptome-wide information at the single-cell level in situ, without the need for cell dissociation. Besides removing any dissociation-related biases in cell type or transcript, integration of transcriptomics and spatial location would create tissue-based atlases of cell types, providing an unbiased version of highly multiplexed in situ hybridization methods [29, 30]. Similarly, other cross-modality technologies are also at various stages of maturation: these include linking single-cell RNA-seq with electrophysiological measurements (Patch-Seq [31]), gene perturbations (CRISPR-Seq and Perturb-Seq [32]), protein expression (CITE-Seq [33]), and lineage tracing (MEMOIR [34], scGESTALT [35]). The large-scale use of all of these technologies, as well as others, is on the horizon, and will result in new multi-modal classification and characterization of cell types in complex tissues. Ultimately, the power of single-cell transcriptomics, and its associated computational methods, will continue to progress as a key component in generating new hypotheses about the organization, regulation, and function of complex tissues. Despite all of these developments, the underlying approach to scRNA-seq data analysis for cell type identification still rests on a basic framework, described below.

## 2   Methods

The following workflow (overview in Fig. 2) describes basic computational steps for identifying molecularly distinct cell types from single-nucleus (sn) RNA-seq data. It does not, however, cover any of the steps relating to the preprocessing, alignment, and quantification of raw sequencing data, which have been described elsewhere [36, 37]. We use the R programming language (https://www.r-project.org), which is a versatile platform for many kinds of genomic analyses, and benefits from the availability of a wide array of statistical and bioinformatic libraries. Over the years, a number of software packages have been developed for single-cell transcritpomic analysis (https://github.com/seandavi/awesome-single-cell), and many of them are available through Bioconductor (https://

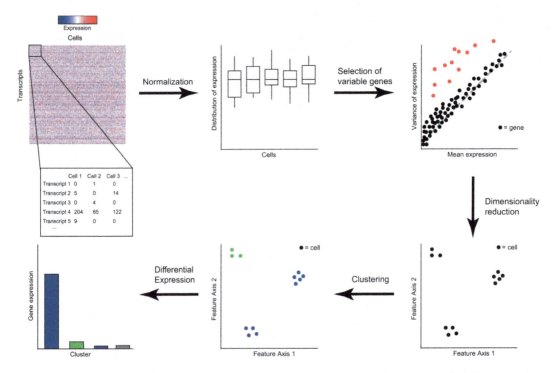

**Fig. 2** Standard computational workflow for identifying transcriptomic types from single-cell RNA-sequencing data, as described in detail in the text, starting from counts tables through to cluster assignment and differential gene expression identification. Although not every computational approach incorporates all of these steps in this order, most involve variations on this set of procedures

www.bioconductor.org/), an open-source archive of bioinformatic R libraries with an active user community. This workflow predominantly uses the Seurat package [38], an actively maintained set of tools for scRNA-seq analysis (https://satijalab.org/seurat/).

Here, we analyze single-nucleus (sn)RNA-seq data covering human frontal cortex (FC), visual cortex (VC), and cerebellum (CB) [39]. While the main text mostly refers to single "cells," the methods below and the general concepts are equally applicable to snRNA-seq data, and also to other single-cell level measurements, such as epigenomic and protein (e.g., mass cytometry) data (although statistical considerations differ).

Our workflow begins with the gene expression matrix **X**, whose rows correspond to genes, and whose columns represent single cells. Entries of the matrix represent digital counts of reads or transcripts, depending on the scRNA-seq protocol that generated the data. Although our presentation employs a specific example dataset, the steps below can be carried out with any gene expression matrix (Fig. 2). The following steps are implemented in RStudio, a free and open source integrated development environment (IDE) for R.

*2.1  Preprocessing: Read the Count Matrix and Setup the Seurat Object*

1. First, we load necessary packages. `utilities.R` is a script that contains some custom functions written by the authors for this workflow                (https://github.com/karthikshekhar/CellTypeMIMB).

```
library(Seurat)
require(ggdendro)
require(Rmisc)
library(Matrix)
library(MASS)
library(xtable)
library(Matrix.utils)
library(DOSE)
library(reshape2)
library(topGO)
library(randomForest)
source("utilities.R")
```

2. We then read in the individual data matrices corresponding to the FC, VC, and CB downloaded from the Gene Expression Omnibus submission of [39] (NCBI Gene Expression Omnibus, GSE97942) [39]. These are stored in a locally accessible folder named `Data`. Since the majority of the entries of these expression matrices are "0," we immediately convert them to the sparse matrix format using the `Matrix` package to reduce the memory footprint.

```
FrontCor_counts = read.table("Data/GSE97930_FrontalCortex_snDrop-seq_UMI_Count
_Matrix_08-01-2017.txt.gz",
    header = TRUE)
FrontCor_counts = Matrix(as.matrix(FrontCor_counts), sparse = TRUE)
VisCor_counts = read.table("Data/GSE97930_VisualCortex_snDrop-seq_UMI_Count_Ma
trix_08-01-2017.txt.gz",
    header = TRUE)
VisCor_counts = Matrix(as.matrix(VisCor_counts), sparse = TRUE)
Cereb_counts = read.table("Data/GSE97930_CerebellarHem_snDrop-seq_UMI_Count_Ma
trix_08-01-2017.txt.gz",
    header = TRUE)
Cereb_counts = Matrix(as.matrix(Cereb_counts), sparse = TRUE)
```

3. Next, we add a "tissue of origin" tag to the three tissue matrices and bind them into a single matrix. The rows of the final matrix correspond to the union of the genes in each of the three tissue matrices. Genes that are missing in any matrix are assumed to not be expressed. We use the `rBind.fill` function in the `Matrix.utils` package to fill in the missing genes,

```
colnames(FrontCor_counts) = paste0("FrontalCortex_", colnames(FrontCor_counts)
)
colnames(VisCor_counts) = paste0("VisualCortex_", colnames(VisCor_counts))
colnames(Cereb_counts) = paste0("Cerebellum_", colnames(Cereb_counts))
Count.mat_sndrop = Matrix.utils::rBind.fill(t(VisCor_counts), t(FrontCor_count
s),
    fill = 0)
Count.mat_sndrop = Matrix.utils::rBind.fill(Count.mat_sndrop, t(Cereb_counts),
fill = 0)
Count.mat_sndrop = t(Count.mat_sndrop)
```

4. Next, we initialize an S4 R object of the class Seurat. The various downstream computations will be performed on this object.

```
# Initialize the Seurat object with the raw (non-normalized data).  Keep all
# genes expressed in >= 10 cells. Keep all cells with at least 500 detected ge
nes
snd <- CreateSeuratObject(raw.data = Count.mat_sndrop, min.cells = 20, min.gen
es = 300,
    project = "snDropBrain")
```

snd@raw.data is a slot in the Seurat object that stores the original gene expression matrix. We can visualize the first 10 rows (genes) and the first 10 columns (cells),

```
snd@raw.data[1:10, 1:10]

## 10 x 10 sparse Matrix of class "dgCMatrix"

##    [[ suppressing 10 column names 'VisualCortex_Ex1_occ17_AAGTGAGTGACC', 'V
isualCortex_Ex1_occ23_CCAGTACGCATC', 'VisualCortex_Ex1_occ24_CCAGGCCTTTCG' ...
]]

##
## A1BG-AS1   . . . . . . . . . .
## A1CF       . . . . . . . . . .
## A2M        . . . . . . . . . .
## A2M-AS1    . . . . . . . . . .
## A2ML1      . 1 . . . . . . . .
## A2ML1-AS1  . . . . . . . 2 . 1
## A2MP1      . . . . . . . . . .
## A3GALT2    . . . . . . . . . .
## A4GALT     . . . . . . . . . .
## AAAS       . . . . . . . . . .
```

5. We then check the dimensions of the normalized expression matrix and the number of cells from each sample. Here snd@ident stores the sample ID's of the cells, corresponding here to their brain region of origin.

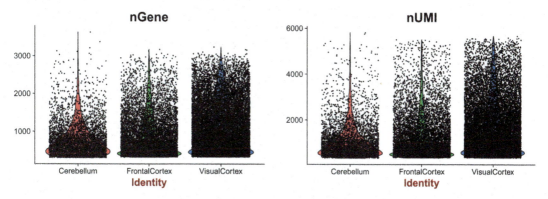

**Fig. 3** Sample-wise (*x*-axis) distribution of the number of genes per cell (Left, *y*-axis) and number of UMIs (i.e., transcripts) per cell (Right, *y*-axis) depicted as violin plots. Dots represent individual cells

```
dim(snd@raw.data)

## [1] 23413 34324

table(snd@ident)
##
##     Cerebellum FrontalCortex   VisualCortex
##           4637         10319          19368
```

6. Thus, we have 23,413 genes and 34,234 cells, with 19,368 cells from the VC, 10,319 cells from the FC and 4637 cells from the CB, respectively. We can visualize common metrics such as number of genes per cell (nGene) and number of transcripts/UMIs per cell (nUMI) as "violin plots" (a fancier version of the good old "box and whisker" plots) using the Seurat plotting command VlnPlot (*see* Fig. 3).

```
VlnPlot(object = snd, features.plot = c("nGene", "nUMI"), nCol = 2, point.size
.use = 0.1)
```

**2.2 Normalize the Data**

1. Because of technical differences in cell-lysis and mRNA capture efficiency, the count vectors of two equivalent cells can differ in the total number of transcripts/UMIs across all genes. This makes it necessary to normalize the data first to attenuate these differences, which is carried out in two steps.

   (a) We rescale the counts in every cell to sum to a constant value. Here, we choose the median of the total transcripts per cell as the scaling factor. This is often referred to as "library-size normalization."

   (b) We apply a logarithmic transformation to the scaled expression values such that $E \leftarrow \log(E + 1)$ (the addition of 1 is to ensure that zeros map to zero values). This transformation has two desirable properties,

- It shrinks values such that the data are more uniformly spread across its range of values, which is especially beneficial if there are outliers.
- Since $\log(A) - \log(B) = \log\left(\frac{A}{B}\right)$, it converts distances along a gene-axis to log-fold change values. This has the consequence that expression differences across cells/samples are treated equally, irrespective of the absolute expression value of the gene. This might be especially desirable for lowly expressed genes such as transcription factors.

```
med_trans = median(Matrix::colSums(snd@raw.data[, snd@cell.names]))
print(med_trans)
snd <- NormalizeData(object = snd, normalization.method = "LogNormalize", scal
e.factor = med_trans)
```

**2.3 Feature Selection: Identify Highly Variable Genes**

1. It is common in analysis of high-dimensional data to choose features that are likely to be informative over features that represent statistical noise, a step known as "Feature Selection." In scRNA-seq data, this is accomplished by choosing genes that are "highly variable" under the assumption that variability in most genes does not represent meaningful biology. An additional challenge is that the level of variability in a gene is related to its mean expression (a phenomenon known as heteroscedacity), which has to be explicitly accounted for. We perform variable gene selection using a recently-published Poisson-Gamma mixture model [40], which was demonstrated to accurately capture the statistical properties of UMI-based scRNA-seq data (Fig. 4).

```
snd = NB.var.genes(snd, do.idents = FALSE, num.sd = 1.2)

## [1] "Identifying variable genes based on UMI Counts. Warning - use this onl
y for UMI based data"
## [1] "Using diffCV = 0.14 as the cutoff"
## [1] "Considering only genes with mean counts less than 3 and more than 0.00
5"
## [1] "Found 1307 variable genes"
print(length(snd@var.genes))

## [1] 1307
```

Thus, we find 1307 variable genes in the data. We refer the reader to other variable gene selection methods, e.g., M3Drop [41], mean-CV regression [42] or Seurat's in-build function `FindVariableGenes`.

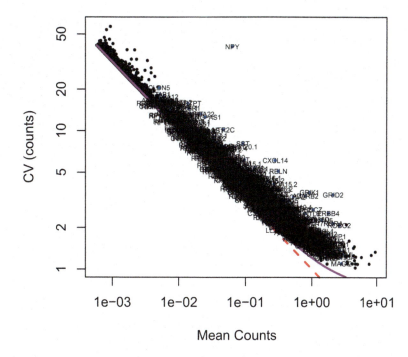

**Fig. 4** Mean (*x*-axis) vs. Coefficient of variation (CV, *y*-axis) of genes (dots). Two null-models of mean-CV relationship—Poisson (dashed-red line) or the Poisson-Gamma mixture model—are also plotted

***2.4 Z-Score the Data and Remove Unwanted Sources of Variation Using Linear Regression***

1. Variation in scRNA-seq data that is relevant to cell identity can be masked by many unwanted sources of variation. A common challenge is batch effects, which can be reflected in both transcriptomic differences and cell-type compositional differences between equivalent experimental batches. As mentioned earlier, variations in lysis efficiency, mRNA capture, and amplification can result in substantial differences between the transcriptomes of equivalent cells. There can be additional sources of variation resulting from biological processes such as cell cycle, response to dissociation, stress, and apoptosis that might dominate the measured transcriptomic state of the cell.

Correcting for such effects continues to be an active area of research, and many sophisticated approaches have been recently introduced [24, 25], but a comprehensive overview is beyond our scope. Here, for demonstrative purposes, we remove variation in gene expression that is highly correlated with library size nUMI. Seurat performs a linear fit to the expression level of every gene using nUMI as a predictor, and returns the residuals as the "corrected" expression values. Next, the expression values are z-scored or standardized along every gene,

$$E_{ij} \leftarrow \frac{E_{ij} - \bar{E}_i}{\sigma_i}$$

Here $E_{ij}$ is the corrected gene expression value of gene $i$ in cell $j$, $\bar{E}_i$ and $\sigma_i$ are the mean and the standard deviation of gene $i$'s expression across all cells. The transformed expression values now have a zero mean and standard deviation equal to 1 across all genes.

2. Removing the effects of nUMI and z-scoring are performed together using Seurat's function ScaleData, which then stores the transformed gene expression values in the slot snd@scale.data.

```
snd <- ScaleData(object = snd, vars.to.regress = c("nUMI"), genes.use = snd@va
r.genes, display.progress = FALSE)
```

### 2.5 The Curse of Dimensionality and Dimensionality Reduction Using PCA

1. Analysis of high-dimensional scRNA-seq data presents numerous challenges, which are often collectively termed the "curse-of-dimensionality" (COD) [43]. For data that is high-dimensional and noisy, samples from the same and different cell subpopulations (i.e., cell types) can appear equidistant from each other, making it difficult to distinguish variability within types and variability across types. Usually COD is dealt with in two ways (Fig. 2). **First,** the number of features/genes can be filtered to only include highly variable genes, as described in the previous section. **Second,** the data can be projected to a lower dimensional subspace using an algorithm that preserves some important properties of the original data, including gene-gene correlations, a choice that is usually informed by the underlying biological question of interest.

There are multiple approaches to dimensionality reduction, such as principal component analysis (PCA) [44], independent component analysis (ICA) [45], non-negative matrix factorization (NMF) [46], autoencoders, and diffusion maps (DM) [47]. Dimensionality reduction results in the compression of raw gene expression data into fewer "composite" variables, each of which is a complex combination of the original gene features, which may be linear or nonlinear depending on the algorithm. These composite features encode the modular structure of the transcriptome alluded to earlier, and may be interpreted as gene modules or "metagenes," with each metagene being defined by a weighted combination of genes. Each cell's observed expression profile can then be interpreted as an aggregate of each metagene weighted by its activity in that particular cell. A situation where multiple metagenes are active in some cells but not others can result in a separation of cells in gene expression space. In this picture, every cell type is a well-separated cloud of points in the reduced dimensional space, whose location is defined by the activity patterns of gene expression modules.

2. Here, we perform Principal Component Analysis (PCA), a classical and extremely versatile dimensionality reduction method that identifies a linear subspace that most accurately captures the variance in the data [44]. Each of the individual axes of this subspace, known as principal vectors (PVs), are linear combinations of the original genes, and the projections of the original data onto these axes are known as principal components (or PCs.)

```
snd <- RunPCA(object = snd, do.print = TRUE, pcs.print = 1:2, genes.print = 5,
pcs.compute = 50)
```

```
## [1] "PC1"
## [1] "KALRN"    "PHACTR1" "PLCB1"    "CHN1"     "KCNQ5"
## [1] ""
## [1] "QKI"      "PLP1"    "CTNNA3" "ST18"    "RNF220"
## [1] ""
## [1] ""
## [1] "PC2"
## [1] "PLP1"     "RNF220"  "MBP"      "MOBP"     "SLC44A1"
## [1] ""
## [1] "ZNF385D" "FSTL5"    "GRID2"    "TIAM1"    "UNC13C"
## [1] ""
## [1] ""
```

Each PV is defined by a set of weights corresponding to the genes (known as the "loadings"). A PV is said to be "driven" by genes with high weights (positive or negative), and two PVs represent independent, orthogonal directions. The printed output of RunPCA lists the genes with the highest magnitude loadings (positive and negative) along the top PVs.

**2.6  Visualize PCA Output**

1. Seurat allows multiple ways to visualize the PCA output, and these are useful to gain biological intuition. VizPCA shows the genes with the highest absolute loadings along any number of user specified PVs (Fig. 5).

```
VizPCA(object = snd, pcs.use = 1:2)
```

2. PCAPlot allows plotting the cells in a reduced dimensional space of PCs, and can often highlight subpopulation structure (Fig. 6).

```
PCAPlot(object = snd, dim.1 = 1, dim.2 = 2, pt.size = 0.4)
```

3. Figures 5 and 6 show that the cells with high values of PC1 are oligodendrocytes, characterized by the high loadings of characteristic genes such as Proteolipid Protein 1 (*PLP1*) and Myelin Basic Protein (*MBP*) (Fig. 5). Next, PCHeatmap allows for

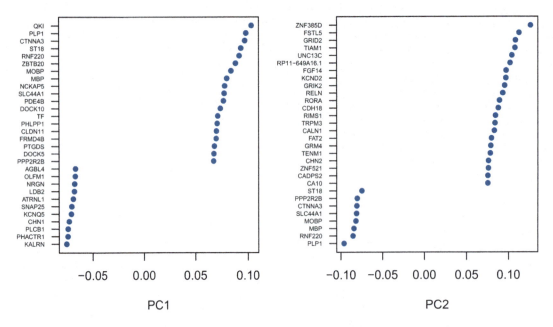

**Fig. 5** Genes (*y*-axis) with the highest negative and positive loadings (*x*-axis) for the top two principal components, PC1 and PC2

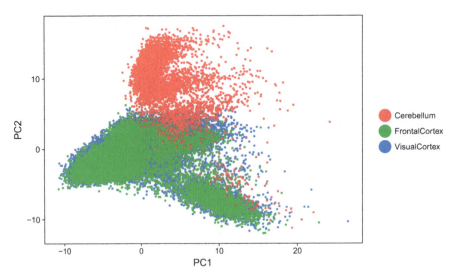

**Fig. 6** Scatter plot showing the scores of individual cells (points) along the top two principal components, PC1 and PC2

easy visualization of the gene expression variation along each PC in the data, and can be particularly useful when trying to decide which PCs to include for further downstream analyses (Fig. 7). Both cells and genes are ordered according to their PCA scores and loadings respectively along each PC. Setting cells.use to a number plots the "extreme" cells on both ends of the spectrum. For example, here we see that genes with

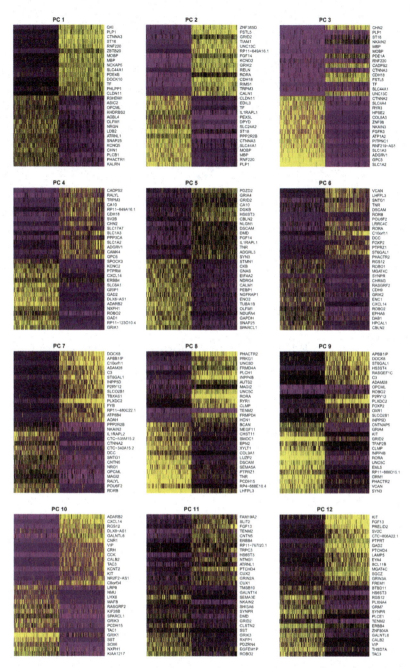

**Fig. 7** Heatmaps showing expression of top 15 positive and negative loading genes in individual cells along PC1–PC12

low values of PC3 are astrocytes, characterized by the expression of the transporters *SLC1A2* and *SLC1A3*.

```
PCHeatmap(object = snd, pc.use = 1:12, cells.use = 500, do.balanced = TRUE, la
bel.columns = FALSE, use.full =
```

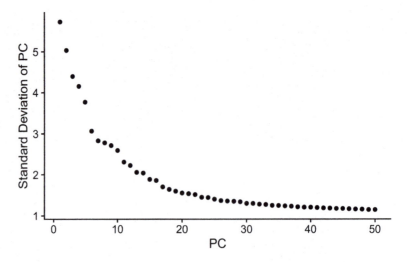

**Fig. 8** Standard-deviation (*y*-axis) accounted for by the top 50 PCs (*x*-axis) to approximately identify the number of significant PCs based on the presence of an "elbow." Approximately 25 PCs are chosen for downstream analysis

While there are many formal methods to determine the number of statistically significant PCs (e.g., *see* Shekhar et al., Cell, 2016 [13]), a particularly easy and popular method is to examine the successive reduction in variance captured by increasing PCs, and identify an "elbow" where inclusion of PCs is of marginal utility (this is often called the "noise floor"). We do this using the Seurat function PCElbowPlot (Fig. 8).

```
PCElbowPlot(object = snd, num.pc = 50)
```

*2.7 Identify Clusters*    1. We choose 25 PCs based on Fig. 8. Every cell in the data is thus reduced from ~23,000 genes to 25 PCs (a ~1000 fold reduction in dimensionality!). Next, we determine subpopulations in this data using Graph-based Clustering [48] using the Seurat FindClusters function. Graph clustering has been widely used in recently scRNA-seq papers and has many desirable properties compared to other methods such as k-means clustering, hierarchical clustering, and density-based clustering. Here, we first build a k-nearest neighbor graph on the data, connecting each cell to its k-nearest neighbor cells based on transcriptional similarity. The nearest neighbors are determined based on proximity in PC space using a Euclidean distance metric. Next, similar to the strategy employed in Levine et al. [49] and Shekhar et al. [13], the graph edge weights are refined based on the Jaccard-similarity metric, which removes spurious edges between clusters. FindClusters implements an algorithm that determines clusters that maximize a mathematical function known as the "modularity" on the Jaccard-weighted k-nearest neighbor graph. The function contains a `resolution` parameter that tunes the granularity of the clustering, with increased values leading to a greater number of clusters. We use a value of 1, but variations in this parameter need to be tested to check for robustness.

```
# save.SNN = T saves the SNN so that the clustering algorithm can be rerun usi
ng
# the same graph but with a different resolution value (see docs for full
# details)
snd <- FindClusters(object = snd, reduction.type = "pca", dims.use = 1:25, res
olution = 1,
    print.output = 0, save.SNN = TRUE, force.recalc = TRUE)
table(snd@ident)
##
##    0     1     2     3     4     5     6     7     8     9    10    11    12    13    14
## 4058  3642  2877  2214  2061  1942  1932  1806  1805  1462  1432  1248  1232   977   931
##   15    16    17    18    19    20    21    22    23    24    25
##  771   673   654   624   599   390   325   285   215   139    30
```

2. Thus, we obtain 26 clusters in the data. We can visualize the clusters using t-distributed stochastic neighbor embedding (t-SNE) [50], a 2-d embedding of the cells that preserves local distances (Fig. 9). The cells are colored according to the cluster labels,

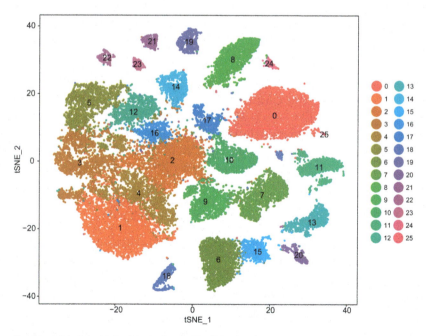

**Fig. 9** Visualization of Lake et al. data using t-distributed neighbor embedding (t-SNE). Cells are colored according to their cluster membership

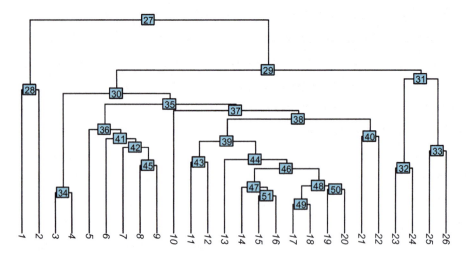

**Fig. 10** Dendrogram showing transcriptional relationships between clusters (nodes)

```
snd <- RunTSNE(object = snd, dims.use = 1:25, do.fast = TRUE)
TSNEPlot(object = snd, do.label = TRUE, pt.size = 0.4)
```

3. Next, we arrange the clusters on a dendrogram based on the similarity of their average transcriptomes using Seurat's `BuildClusterTree` function (Fig. 10). This helps in visualizing relationships between clusters, and also reveals subgroups of related clusters.

```
snd <- BuildClusterTree(snd, do.reorder = T, reorder.numeric = T, genes.use =
snd@var.genes, show.progress = FALSE)
```

4. At this point, it is important to note that whether or not we have found the "optimal" number of clusters is open to interpretation. Importantly, the criterion of what constitutes a cell type cluster must be independent of the algorithm's objective—it could be data driven, such as a minimum number of differentially expressed genes enriched in that cluster compared to the rest, or the ability of the algorithm to recover certain well-known types (i.e., ground truth). Often, however, the validation of scRNA-seq clusters requires aligning molecular identity to other cell modalities such as morphology, location, and function through experimental techniques.

Here we adopt a data-driven criterion to assess cluster stability. Briefly, Seurat's `AssessNode` function trains a classifier on each binary node of the dendrogram, and calculates the classification error for left/right clusters. We can use this information to collapse any node that exhibits >15% classification error.

```
node.scores <- AssessNodes(snd)

## Growing trees.. Progress: 74%. Estimated remaining time: 10 seconds.
## Growing trees.. Progress: 49%. Estimated remaining time: 32 seconds.
## Growing trees.. Progress: 73%. Estimated remaining time: 11 seconds.
## Growing trees.. Progress: 98%. Estimated remaining time: 0 seconds.

node.scores <- node.scores[order(node.scores$oobe, decreasing = T), ]
print(head(node.scores))

##      node        oobe
## 20     49 0.11187095
## 19     48 0.09380741
## 21     50 0.07690141
## 16     46 0.07390206
## 17     47 0.05475130
## 5      34 0.03921569
```

### 2.8 Compare Clusters with Original Cell Type Labels from Lake et al. [39]

1. Here, we see that the maximum "out of bag classification error" (OOBE), is less than our threshold. Thus, we retain all 26 clusters. Next, we compare our clustering result to the cluster labels published in Lake et al. [39], which nominated 33 clusters in their analysis. While we have obviously fewer clusters, it would be interesting to examine how they compare to Lake et al.'s results. We first read in their cluster labels,

```
lake_clusters = unlist(lapply(strsplit(colnames(snd@data), "_"), function(x) x
[2]))
names(lake_clusters) = colnames(snd@data)
length(table(lake_clusters))

## [1] 33

head(table(lake_clusters))

## lake_clusters
##  Ast   End   Ex1   Ex2  Ex3a  Ex3b
## 2409   218  5669   310   588  1089
```

Here, Ast refers to astrocytes, End refers to endothelial cells, Ex1 refers to Excitatory neuron group 1, and so on. To compare our cluster labels against Lake et al.'s, we plot a "confusion matrix," where each row corresponds to one of Lake et al's 33 clusters, while each column corresponds to our cluster (Fig. 11). The matrix is row-normalized to depict how each cluster of Lake et al. distributes across our clusters.

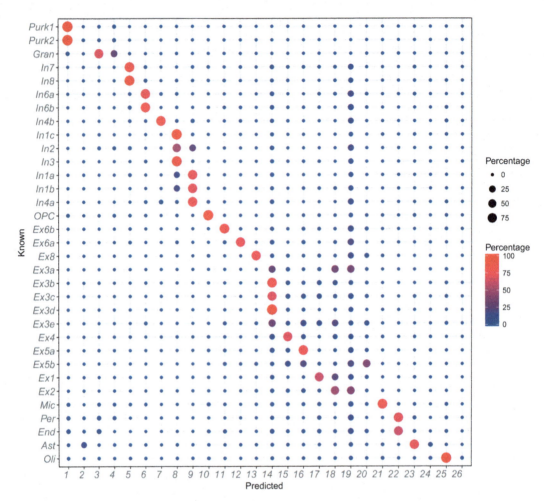

**Fig. 11** Transcriptional correspondence between clusters determined from the Lake et al. dataset in this study and in the original study. Circles depict the percentage of cells of a given Lake et al. cluster (row) assigned to a cluster determined above (column)

```
A = table(lake_clusters, snd@ident)
# we post-hoc specify the row order to make the matrix diagonal
row.order = c("Purk1", "Purk2", "Gran", "In7", "In8", "In6a", "In6b", "In4b",
"In1c",
    "In2", "In3", "In1a", "In1b", "In4a", "OPC", "Ex6b", "Ex6a", "Ex8", "Ex3a"
, "Ex3b",
    "Ex3c", "Ex3d", "Ex3e", "Ex4", "Ex5a", "Ex5b", "Ex1", "Ex2", "Mic", "Per",
"End",
    "Ast", "Oli")
a = plotConfusionMatrix(A[row.order, ])
```

2. Encouragingly, we see that although our analysis workflow was agnostic to the results reported in the original paper, many of our clusters exhibit a 1:1 correspondence with the clusters of Lake et al. For example, Cluster 21 ($n = 624$) corresponds to

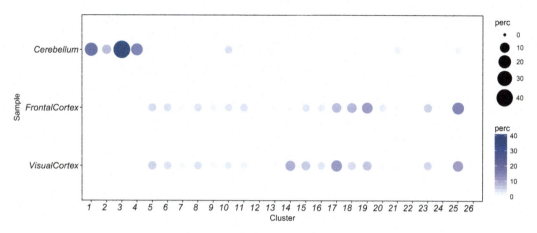

**Fig. 12** Cluster composition of each brain region. Circles indicate the proportion of each cluster (columns) within each region (row). Each row sums to 1

Microglia (Mic), while Cluster 25 ($n = 4058$ cells) corresponds to Oligodendrocytes (Oli). In cases where multiple Lake et al. clusters map to our clusters, these are related. For example, Purkinje cell clusters Purk1 and Purk2 map to Cluster 1 ($n = 977$), while inhibitory neurons In6a and In6b map to Cluster 6 ($n = 1462$). It is likely that a second round of iterative clustering might be necessary to resolve differences between closely related types such as In6a and In6b. While all of this is encouraging, we also note some discrepancies—Clusters 2 ($n = 390$), 24 ($n = 139$) and 26 ($n = 30$), do not really correspond to any of the Lake et al., clusters, while clusters 18 ($n = 2061$) and 19 ($n = 2877$) appear to nonspecifically map to many Lake et al. clusters.

3. We can visualize the cluster composition of each of the three brain regions (Fig. 12),

```
# Each row is normalized to a 100%
plot_sample_dist(snd, row.scale = TRUE)
```

As can be seen Clusters 1–4 and 26, which include Purkinje neurons and Cerebellar granule cells, are exclusive to the CB sample, while majority of the remaining clusters are derived from the FC and VC samples.

*2.9 Identify Cluster-Specific Differentially Expressed Genes*

1. Next, we find cluster-specific markers by performing a differential expression (DE) analysis between each cluster and the rest using Seurat's FindMarkers function. FindMarkers supports the use of multiple statistical approaches for DE (specified in the test.use parameter, *see* Seurat documentation). Here, we use the Student's t-test, as it is computationally efficient. However, we note that there are many limitations to using the t-test for single-cell RNA-seq data, particularly its

inability to account for zero inflation. Readers must explore other methods such as MAST and tweeDEseq supported by Seurat (for a comprehensive review on DE methods, *see* Soneson and Robinson [51]).

```
# find markers for every cluster compared to all remaining cells, report only the
# positive ones
snd.markers <- FindAllMarkers(object = snd, only.pos = TRUE, min.pct = 0.25, t
hresh.use = 0.25,
    test.use = "t", max.cells.per.ident = 1000)
head(snd.markers)
```

```
##                  p_val avg_logFC pct.1 pct.2     p_val_adj cluster   gene
## GRID2   0.000000e+00  2.283338 0.999 0.422  0.000000e+00       1  GRID2
## RORA    0.000000e+00  1.760396 0.991 0.588  0.000000e+00       1   RORA
## INPP4B 4.623397e-308  1.120327 0.878 0.228 1.082476e-303       1 INPP4B
## GRM1   5.438182e-270  0.964801 0.831 0.164 1.273242e-265       1   GRM1
## UNC5C  5.279446e-252  1.026785 0.889 0.339 1.236077e-247       1  UNC5C
## SYN3   1.012546e-197  0.825126 0.820 0.308 2.370674e-193       1   SYN3
```

2. The output is a data.frame object summarizing the cluster-specific markers. Here, each row is a gene that is enriched in a cluster indicated in the column cluster. pct.1 is the proportion of cells in the cluster that express this marker, while pct.2 is the proportion of cells in the background that express this marker. We can examine markers for a given cluster as follows,

```
clust = 25   # Oligodendrocytes
head(subset(snd.markers, cluster == clust & pct.2 < 0.1))
```

```
##                  p_val avg_logFC pct.1 pct.2     p_val_adj cluster   gene
## PLP11   0.000000e+00 1.6572808 0.873 0.084  0.000000e+00      25   PLP1
## MOBP1  2.972597e-177 1.0967866 0.660 0.057 6.959740e-173      25   MOBP
## TF     7.315453e-133 0.8444486 0.513 0.039 1.712767e-128      25     TF
## FRMD4B1 2.760596e-125 0.7469469 0.519 0.084 6.463384e-121      25 FRMD4B
## CLDN11 1.497425e-117 0.7452380 0.483 0.051 3.505920e-113      25 CLDN11
## TMEM144 4.327291e-105 0.6684642 0.434 0.048 1.013149e-100      25 TMEM144
```

3. As expected, the top two genes are *PLP1* (Proteolipid Protein 1) and *MOBP* (Myelin-Associated Oligodendrocyte Basic Protein), classical markers of Oligodendrocytes. Next, we examine cluster 12 (an excitatory neuronal cluster), which corresponds to Ex6a, and is marked by multiple genes including *HTR2C* and *NPSR1-AS1* (Fig. 13).

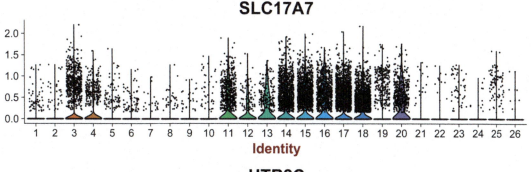

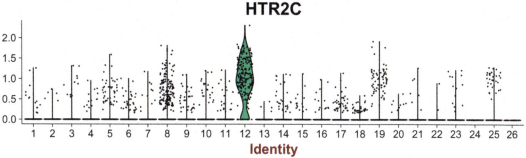

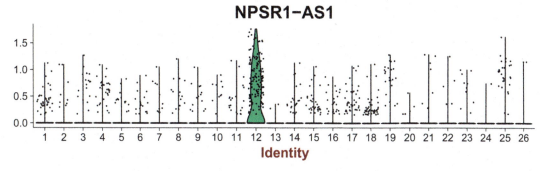

**Fig. 13** Cluster-expression of a pan-excitatory neuronal marker *SLC17A7* (top) and markers specific to cluster 12 *HTR2C* (middle) and *NPSR1-AS1* (bottom)

```
head(subset(snd.markers, cluster == 12 & pct.2 < 0.1))

##                      p_val avg_logFC pct.1 pct.2   p_val_adj cluster
## HTR2C        2.715453e-67 1.0524759 0.842 0.018 6.357689e-63      12
## RP11-420N3.3 4.584219e-39 0.6270714 0.702 0.095 1.073303e-34      12
## NPSR1-AS1    4.002079e-37 0.6707752 0.651 0.013 9.370068e-33      12
## IFNG-AS1     2.622851e-35 0.6828199 0.623 0.011 6.140881e-31      12
## PCP4         1.398067e-22 0.4261675 0.535 0.080 3.273294e-18      12
## CRYM         1.012721e-21 0.3684937 0.507 0.068 2.371085e-17      12
##                      gene
## HTR2C               HTR2C
## RP11-420N3.3 RP11-420N3.3
## NPSR1-AS1       NPSR1-AS1
## IFNG-AS1         IFNG-AS1
## PCP4                 PCP4
## CRYM                 CRYM

VlnPlot(snd, c("SLC17A7", "HTR2C", "NPSR1-AS1"), nCol = 1, point.size.use = 0.
01)
```

Examining the identity of these clusters in detail is beyond the scope of this workflow. Readers are encouraged dig deeper, and attempt to test variations in the methods outlined above. We end by demonstrating two common approaches to interpret results: (a) Examining gene-set enrichments, and (b) aligning clusters to alternative datasets.

*2.10 Examine Clusters for Enrichment of Biological Processes*

1. After identifying markers, we can evaluate whether cluster-specific genes are enriched for any Gene Ontology (GO), Disease Ontology (DO), or Disease Gene Network (DGN) gene lists or categories. Each of these calls has multiple parameters, reflecting stringency of statistical overlap, but they are useful tools to evaluate clusters for functional or disease relevance.

```
#### evaluate for each cluster###
require(org.Hs.eg.db)

## Loading required package: org.Hs.eg.db

##

x = as.list(org.Hs.egALIAS2EG)
geneList = rep(0, nrow(Count.mat_sndrop))
names(geneList) = rownames(Count.mat_sndrop)
geneList = geneList[intersect(names(geneList), names(x))]
newallgenes = names(geneList)
for (ii in 1:length(geneList)) {
    names(geneList)[ii] = x[[names(geneList)[ii]]][1]
}

gene_enrichment_results = list()

for (cl in as.character(unique(snd.markers$cluster))) {
    print(paste0("Running cluster ", cl))
    testgenes = subset(snd.markers, cluster == cl)$gene
    gene_enrichment_results[[cl]] = list()
    #### Run against topGO####
    testgeneList = geneList
    testgeneList[which(newallgenes %in% testgenes)] = 1
    genegene_enrichment_results = list()
    tab1 = c()
    for (ont in c("BP", "MF", "CC")) {
        sampleGOdata <- suppressMessages(new("topGOdata", description = "Simpl
e session",
            ontology = ont, allGenes = as.factor(testgeneList), nodeSize = 10,
annot = annFUN.org,
            mapping = "org.Hs.eg.db", ID = "entrez"))
        resultTopGO.elim <- suppressMessages(runTest(sampleGOdata, algorithm =
"elim",
            statistic = "Fisher"))
        resultTopGO.classic <- suppressMessages(runTest(sampleGOdata, algorith
```

```
m = "classic",
            statistic = "Fisher"))
        ## look at results
        tab1 <- rbind(tab1, GenTable(sampleGOdata, Fisher.elim = resultTopGO.e
lim,
            Fisher.classic = resultTopGO.classic, orderBy = "Fisher.elim", top
Nodes = 200))
    }
    gene_enrichment_results[[cl]][["topGO"]] = tab1

    #### Run against DOSE####
    x <- suppressMessages(enrichDO(gene = names(testgeneList)[testgeneList ==
1],
        ont = "DO", pvalueCutoff = 1, pAdjustMethod = "BH", universe = names(t
estgeneList),
        minGSSize = 5, maxGSSize = 500, qvalueCutoff = 1, readable = T))
    gene_enrichment_results[[cl]][["DO"]] = x
    dgn <- suppressMessages(enrichDGN(names(testgeneList)[testgeneList == 1]))
    gene_enrichment_results[[cl]][["DGN"]] = dgn
}

## [1] "Running cluster 1"
## [1] "Running cluster 2"
## [1] "Running cluster 3"
## [1] "Running cluster 4"
## [1] "Running cluster 5"
## [1] "Running cluster 6"
## [1] "Running cluster 7"
## [1] "Running cluster 8"
## [1] "Running cluster 9"
## [1] "Running cluster 10"
## [1] "Running cluster 11"
## [1] "Running cluster 12"
## [1] "Running cluster 13"
## [1] "Running cluster 14"
## [1] "Running cluster 15"
## [1] "Running cluster 16"
## [1] "Running cluster 17"
## [1] "Running cluster 18"
## [1] "Running cluster 19"
## [1] "Running cluster 20"
## [1] "Running cluster 21"
## [1] "Running cluster 22"
## [1] "Running cluster 23"
## [1] "Running cluster 24"
## [1] "Running cluster 25"
## [1] "Running cluster 26"

save(gene_enrichment_results, file = "gene_enrichment_analysis.rda")
```

2. As an example, view the GO, DO, and DGN categories enriched for genes distinguishing cluster 1 (Purkinje Neurons). Note that the categories are arranged by adjusted p-value, and many are not significantly enriched.

```
gene_enrichment_results[["1"]][["topGO"]][1:5, ]

##        GO.ID                                          Term Annotated
## 1 GO:0035235 ionotropic glutamate receptor signaling ...        25
## 2 GO:0071625                          vocalization behavior        17
## 3 GO:0007612                                       learning       131
## 4 GO:1904861                     excitatory synapse assembly        10
## 5 GO:0051965        positive regulation of synapse assembly        63
##   Significant Expected Rank in Fisher.classic Fisher.elim Fisher.classic
## 1           6     0.14                       6     4.1e-09         5.1e-09
## 2           4     0.09                      16     2.0e-06         2.3e-06
## 3           7     0.73                      24     8.4e-06         1.1e-05
## 4           3     0.06                      31     2.0e-05         2.2e-05
## 5           5     0.35                      32     2.6e-05         3.1e-05


gene_enrichment_results[["1"]][["DO"]][1:5, ]

##                            ID                           Description
## DOID:0060037 DOID:0060037 developmental disorder of mental health
## DOID:1827         DOID:1827         idiopathic generalized epilepsy
## DOID:0060041 DOID:0060041                 autism spectrum disorder
## DOID:12849       DOID:12849                         autistic disorder
## DOID:0060040 DOID:0060040       pervasive developmental disorder
##              GeneRatio  BgRatio       pvalue     p.adjust       qvalue
## DOID:0060037     12/57 344/7041 1.474101e-05 0.002046357 0.001921672
## DOID:1827         4/57  20/7041 1.698221e-05 0.002046357 0.001921672
## DOID:0060041      7/57 171/7041 4.187781e-04 0.025231379 0.023694023
## DOID:12849        7/57 171/7041 4.187781e-04 0.025231379 0.023694023
## DOID:0060040      7/57 180/7041 5.704404e-04 0.027495225 0.025819932
##
geneID
## DOID:0060037 AUTS2/CACNA1A/CNTNAP5/GRIA3/MACROD2/NBEA/NLGN1/NOS1AP/NRXN1/SH
ANK2/SYN3/XKR4
## DOID:1827                                                       CACNA1A/GA
D2/HCN1/KCNMA1
## DOID:0060041                          AUTS2/CNTNAP5/MACROD2/NBEA/NLGN
1/NOS1AP/NRXN1
## DOID:12849                           AUTS2/CNTNAP5/MACROD2/NBEA/NLGN
1/NOS1AP/NRXN1
## DOID:0060040                         AUTS2/CNTNAP5/MACROD2/NBEA/NLGN
1/NOS1AP/NRXN1
##              Count
## DOID:0060037    12
## DOID:1827        4
## DOID:0060041     7
## DOID:12849       7
## DOID:0060040     7


gene_enrichment_results[["1"]][["DGN"]][1:5, ]

##                           ID               Description GeneRatio
## umls:C1272641 umls:C1272641  Systemic arterial pressure     19/89
## umls:C1271104 umls:C1271104        Blood pressure finding     18/89
## umls:C0236969 umls:C0236969 Substance-Related Disorders     10/89
## umls:C1510586 umls:C1510586   Autism Spectrum Disorders     10/89
## umls:C0007758 umls:C0007758            Cerebellar Ataxia      6/89
```

```
##                    BgRatio      pvalue      p.adjust        qvalue
## umls:C1272641 442/17381 7.549992e-13 3.439175e-10 3.250238e-10
## umls:C1271104 386/17381 8.367822e-13 3.439175e-10 3.250238e-10
## umls:C0236969 167/17381 1.354158e-08 3.710394e-06 3.506557e-06
## umls:C1510586 246/17381 5.129828e-07 1.054180e-04 9.962667e-05
## umls:C0007758 107/17381 1.815118e-05 2.984054e-03 2.820120e-03
##
geneID
## umls:C1272641 105/26053/129684/64478/728215/9758/442117/23072/3778/140733/2
3026/30010/5592/6446/57419/23345/84216/440279/114786
## umls:C1271104        105/26053/129684/64478/728215/9758/442117/23072/3778/140
733/23026/30010/5592/57419/23345/84216/440279/114786
## umls:C0236969                                                          105/
3899/64478/55691/2893/10207/140733/9369/23345/114786
## umls:C1510586                                                          260
53/776/1804/140733/26960/22871/9378/9369/22941/26137
## umls:C0007758
773/2259/2572/2895/2911/23345
##               Count
## umls:C1272641    19
## umls:C1271104    18
## umls:C0236969    10
## umls:C1510586    10
## umls:C0007758     6
```

## 2.11  Compare with Mouse Cortical Cell Types

1. One of the many challenges in cell type classification studies is that of aligning clusters across different datasets, which might include different batches, different conditions (e.g., normal vs. disease), or even different species. Here we attempt to map clusters from a dataset of visual cortex (VC) neurons isolated and profiled from adult mouse using the Smart-seq method [15] to our Human CB, VC, and FC clusters using a supervised learning algorithm. We use a multiclass classification approach described previously [13].

First, we read in the mouse VC data comprised of 1679 cells and create a *Seurat* S4 object. To match the gene ID's to Human data, we capitalize all gene names—note that a more exact, albeit lengthier approach, would be to match genes based on an appropriate orthology database. We also read in the cluster assignments of each cell. Tasic et al. identified 49 transcriptomic types, comprising 23 inhibitory, 19 excitatory, and 7 non-neuronal types [15]. We next select features to train our classifier. We identify variable genes using Seurat's `FindVariableGenes` function (Fig. 14), which is more appropriate for Smart-seq data [40]. After expanding the set of variable genes in the snRNA-seq data using `NB.var.genes`, we compute the common variable genes to train a multi-class classifier.

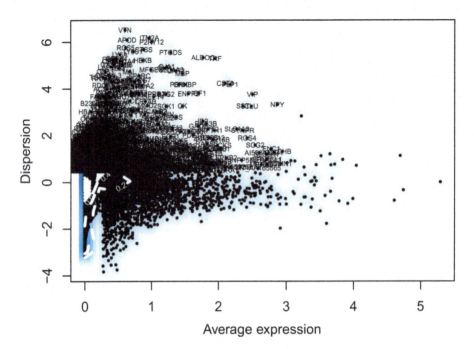

**Fig. 14** Identification of highly variable genes using Seurat's `FindVariableGenes` function (see documentation for details). This is an appropriate strategy for feature selection on scRNA-seq that does not contain UMIs

```
tasic_data = Matrix(as.matrix(read.csv("tasic_2016/genes_counts_2.csv", header
= TRUE,
    row.names = 1)), sparse = TRUE)
# Change gene name format to match to Human
rownames(tasic_data) = toupper(rownames(tasic_data))
mouse_vc <- CreateSeuratObject(raw.data = tasic_data, min.cells = 10, min.gene
s = 500,
    project = "tasic")
mouse.clusters = read.csv("tasic_2016/cluster_assignment_simple.csv", row.name
s = 2)
mouse.labels = mouse.clusters$primary
names(mouse.labels) = rownames(mouse.clusters)
mouse_vc@meta.data$type = mouse.labels[rownames(mouse_vc@meta.data)]
mouse_vc = SetAllIdent(mouse_vc, id = "type")
mouse_vc <- NormalizeData(object = mouse_vc, normalization.method = "LogNormal
ize",
    scale.factor = 10000)
mouse_vc <- FindVariableGenes(object = mouse_vc, mean.function = ExpMean, disp
ersion.function = LogVMR, x.low.cutoff = 0.0125, x.high.cutoff = 3, y.cutoff =
0.5, set.var.genes = TRUE)
```

```
var.genes_snd = NB.var.genes(snd, do.idents = FALSE, num.sd = 0.8, do.plot = F
ALSE, set.var.genes = FALSE)

## [1] "Identifying variable genes based on UMI Counts. Warning - use this onl
y for UMI based data"
## [1] "Using diffCV = 0.11 as the cutoff"
## [1] "Considering only genes with mean counts less than 3 and more than 0.00
5"
## [1] "Found 1977 variable genes"

var.genes = intersect(var.genes_snd, mouse_vc@var.genes)
```

2. Next, we train a Random Forest (RF) model [52] on the snRNA-seq data and use that to assign cluster labels to mouse VC data. Given a cell, the classifier maps it to one of 26 clusters. To account for scale differences between the snRNA-seq (3′-biased, UMI-based) and Smart-seq (full-length, non-UMI-based), we standardize the two datasets (z-score values along each gene). After training it on the snRNA-seq data, we apply this classifier to each cell from the mouse VC data, and assign it to one of 26 snRNA-seq clusters.

```
rf_model = RF_train(snd, var.genes, do.scale = TRUE)

## [1] "Using mininum of 50 percent cells or 700 cells per training"
g"
## [1] 13067

pred.labels <- predict(rf_model, t(t(scale(t(mouse_vc@data[var.genes, ])))))
pred.labels <- factor(as.numeric(as.character(pred.labels)) + 1)
names(pred.labels) = colnames(mouse_vc@data)
```

3. How do the cluster assignments compare with the cluster labels obtained from Tasic et al. [15]? Note that the latter labels were not used in any way to either construct the classifier, or to influence the cluster assignment of cells. It would therefore be interesting to see if there is any correspondence between mouse cortical cell types, and their assigned "Human" type based on an unbiased classifier. We examine the confusion matrix, as before (Fig. 15),

```
a = plotConfusionMatrix(table(mouse.labels, pred.labels), order = "Row")
```

The rows correspond to the Tasic et al. clusters, while each column corresponds to an snRNA-seq cluster. The matrix is row-normalized, such that each row adds up to a 100%. First, we see that Clusters 1–4 and 26, which are of Cerebellar origin, receive very few matches from mouse VC data, which largely map to Human clusters originating from VC and FC samples. Among non-neuronal cells, we see that mouse astrocytes and oligodendrocytes map to clusters 23 and 25, which are Human astrocytes and oligodendrocytes, respectively. Inhibitory neuronal groups expressing parvalbumin (*Pvalb*), Somatostatin (*Sst*), and Vasoactive intestinal peptide (*Vip*) map to clusters 6, 5, and 8 respectively. Examining the expression of these markers in the snRNA-seq data validates the RF cluster assignments (Fig. 16). Thus, despite the fact that these two data sets differ in species (human vs. mouse), cell fraction profiled (cytoplasmic vs. nucleus-only), profiling method (Smart-Seq vs. droplet-based sequencing), and clustering method (gene clustering vs. PCA-based methods vs. PCA-Louvain clustering), the overall results are comparable and interpretable, suggesting that the transcriptomic space these cells occupy is being appropriately parsed into subtypes.

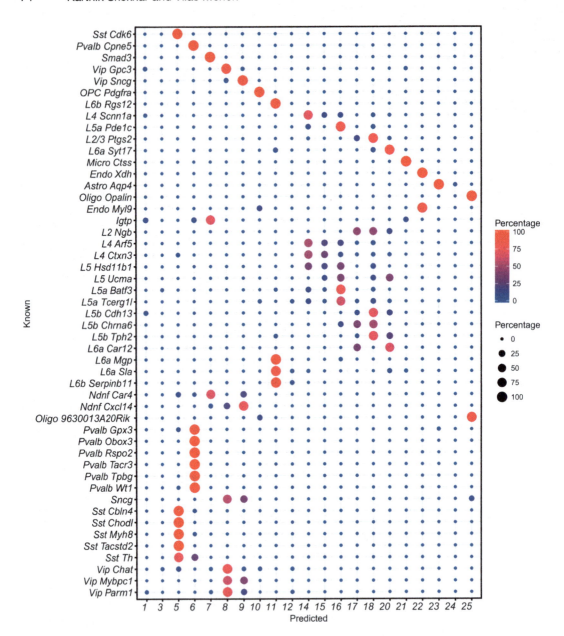

**Fig. 15** Transcriptional correspondence between mouse cortical clusters reported in Tasic et al. [15] (rows) and those in this study (columns). Representation as in Fig. 11

```
VlnPlot(snd, c("PVALB", "SST", "VIP"), nCol = 1, point.size.use = 0.01)
```

This concludes the basic workflow. We can save files from the analysis as follows.

```
save(list=c("snd","snd.markers","rf_model","mouse_vc"), file="LakeSeuratAnalys
isObject.Rdata")
```

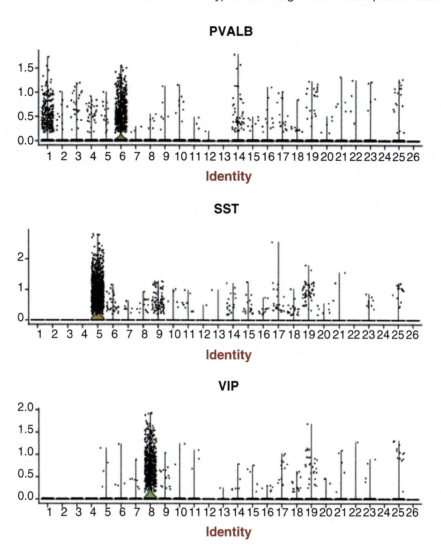

**Fig. 16** Expression of three classic markers known to distinguish inhibitory neuronal categories *PVALB* (top), *SST* (middle), and *VIP* (bottom)

## Acknowledgments

K. S. would like to acknowledge support from NIH 1K99EY028625-01, the Klarman Cell Observatory, and the laboratory of Dr. Aviv Regev at the Broad Institute. We would like to gratefully acknowledge critical feedback from Drs. Inbal Benhar and Jose Ordovas-Montanes.

## References

1. Vickaryous MK, Hall BK (2006) Human cell type diversity, evolution, development, and classification with special reference to cells derived from the neural crest. Biol Rev Camb Philos Soc 81(3):425–455

2. Regev A et al (2017) The human cell atlas. Elife:6

3. Tosches MA et al (2018) Evolution of pallium, hippocampus, and cortical cell types revealed by single-cell transcriptomics in reptiles. Science 360(6391):881–888

4. Boisset JC et al (2018) Mapping the physical network of cellular interactions. Nat Methods

5. Tanay A, Regev A (2017) Scaling single-cell genomics from phenomenology to mechanism. Nature 541(7637):331–338

6. Trapnell C (2015) Defining cell types and states with single-cell genomics. Genome Res 25(10):1491–1498

7. Cleary B et al (2017) Efficient generation of transcriptomic profiles by random composite measurements. Cell 171(6):1424–1436.e18

8. Klein AM et al (2015) Droplet barcoding for single-cell transcriptomics applied to embryonic stem cells. Cell 161(5):1187–1201

9. Macosko EZ et al (2015) Highly parallel genome-wide expression profiling of individual cells using nanoliter droplets. Cell 161(5):1202–1214

10. Zheng GX et al (2017) Massively parallel digital transcriptional profiling of single cells. Nat Commun 8:14049

11. Habib N et al (2016) Div-Seq: single-nucleus RNA-Seq reveals dynamics of rare adult newborn neurons. Science 353(6302):925–928

12. Lake BB et al (2016) Neuronal subtypes and diversity revealed by single-nucleus RNA sequencing of the human brain. Science 352(6293):1586–1590

13. Shekhar K et al (2016) Comprehensive classification of retinal bipolar neurons by single-cell transcriptomics. Cell 166(5):1308–1323.e30

14. Villani A-C et al (2017) Single-cell RNA-seq reveals new types of human blood dendritic cells, monocytes, and progenitors. Science 356(6335):eaah4573

15. Tasic B et al (2016) Adult mouse cortical cell taxonomy revealed by single cell transcriptomics. Nat Neurosci 19(2):335–346

16. Zeng H, Sanes JR (2017) Neuronal cell-type classification: challenges, opportunities and the path forward. Nat Rev Neurosci 18(9):530

17. Stegle O, Teichmann SA, Marioni JC (2015) Computational and analytical challenges in single-cell transcriptomics. Nat Rev Genet 16(3):133

18. Arendt D (2008) The evolution of cell types in animals: emerging principles from molecular studies. Nat Rev Genet 9(11):868–882

19. Ecker JR et al (2017) The BRAIN initiative cell census consortium: lessons learned toward generating a comprehensive BRAIN cell atlas. Neuron 96(3):542–557

20. Kolodziejczyk AA et al (2015) The technology and biology of single-cell RNA sequencing. Mol Cell 58(4):610–620

21. Islam S et al (2014) Quantitative single-cell RNA-seq with unique molecular identifiers. Nat Methods 11(2):163

22. Menon V (2017) Clustering single cells: a review of approaches on high- and low-depth single-cell RNA-seq data. Brief Funct Genomics

23. Hicks SC, Teng M, Irizarry RA (2015, 025528) On the widespread and critical impact of systematic bias and batch effects in single-cell RNA-Seq data. bioRxiv

24. Butler A et al (2018) Integrating single-cell transcriptomic data across different conditions, technologies, and species. Nat Biotechnol 36(5):411

25. Haghverdi L et al (2018) Batch effects in single-cell RNA-sequencing data are corrected by matching mutual nearest neighbors. Nat Biotechnol 36:421–427

26. Lopez R et al (2018) Bayesian inference for a generative model of transcriptome profiles from single-cell RNA sequencing. bioRxiv:292037

27. Lee JH et al (2014) Highly multiplexed subcellular RNA sequencing in situ. Science 343(6177):1360–1363

28. Stahl PL et al (2016) Visualization and analysis of gene expression in tissue sections by spatial transcriptomics. Science 353(6294):78–82

29. Chen KH et al (2015) Spatially resolved, highly multiplexed RNA profiling in single cells. Science 348(6233):aaa6090

30. Lubeck E et al (2014) Single-cell in situ RNA profiling by sequential hybridization. Nat Methods 11(4):360

31. Fuzik J et al (2016) Integration of electrophysiological recordings with single-cell RNA-seq data identifies neuronal subtypes. Nat Biotechnol 34(2):175

32. Dixit A et al (2016) Perturb-Seq: dissecting molecular circuits with scalable single-cell

RNA profiling of pooled genetic screens. Cell 167(7):1853–1866.e17

33. Stoeckius M et al (2017) Simultaneous epitope and transcriptome measurement in single cells. Nat Methods 14(9):865

34. Frieda KL et al (2017) Synthetic recording and in situ readout of lineage information in single cells. Nature 541(7635):107–111

35. Raj B et al (2018) Simultaneous single-cell profiling of lineages and cell types in the vertebrate brain. Nat Biotechnol 36(5):442–450

36. Pertea M et al (2016) Transcript-level expression analysis of RNA-seq experiments with HISAT, StringTie and Ballgown. Nat Protoc 11(9):1650

37. Villani AC, Shekhar K (2017) Single-cell RNA sequencing of human T cells. Methods Mol Biol 1514:203–239

38. Satija R et al (2015) Spatial reconstruction of single-cell gene expression data. Nat Biotechnol 33(5):495–502

39. Lake BB et al (2018) Integrative single-cell analysis of transcriptional and epigenetic states in the human adult brain. Nat Biotechnol 36 (1):70–80

40. Pandey S et al (2018) Comprehensive identification and spatial mapping of Habenular neuronal types using single-cell RNA-Seq. Curr Biol 28(7):1052–1065.e7

41. Andrews TS, Hemberg M (2017) Identifying cell populations with scRNASeq. Mol Asp Med

42. Brennecke P et al (2013) Accounting for technical noise in single-cell RNA-seq experiments. Nat Methods 10(11):1093

43. Keogh E, Mueen A (2017) Curse of dimensionality. In: Encyclopedia of machine learning and data mining. Springer, pp 314–315

44. Hotelling H (1933) Analysis of a complex of statistical variables into principal components. J Educ Psychol 24(6):417

45. Hyvärinen A, Karhunen J, Oja E (2004) Independent component analysis, vol 46. Wiley, New York

46. Lee DD, Seung HS (2001) Algorithms for non-negative matrix factorization. In: Leen TK, Dietterich TG, Tresp V (eds) Advances in neural information processing systems, vol 13. MIT, Cambridge, UK

47. Haghverdi L et al (2016) Diffusion pseudo-time robustly reconstructs lineage branching. Nat Methods 13(10):845

48. Lancichinetti A, Fortunato S (2009) Community detection algorithms: a comparative analysis. Phys Rev E Stat Nonlinear Soft Matter Phys 80(5 Pt 2):056117

49. Levine JH et al (2015) Data-driven phenotypic dissection of AML reveals progenitor-like cells that correlate with prognosis. Cell 162 (1):184–197

50. LVD M, Hinton G (2008) Visualizing data using t-SNE. J Mach Learn Res 9 (Nov):2579–2605

51. Soneson C, Robinson MD (2018) Bias, robustness and scalability in single-cell differential expression analysis. Nat Methods 15 (4):255

52. Breiman L (2001) Random forests. Mach Learn 45(1):5–32

# Chapter 5

# Rare Cell Type Detection

## Lan Jiang

## Abstract

High-throughput single-cell technologies have great potential to discover new cell types. Here, we present a novel computational method, called GiniClust (Jiang et al., Genome Biol 17(1):144, 2016), to overcome the challenge of detecting rare cell types that are distinct from a large population.

**Key words** Clustering, Single-cell analysis, RNA-seq, qPCR, Gini index, Rare cell type

## 1 Introduction

The cell is the basic unit of structure and function in life; however, our knowledge of cell types remains largely incomplete. Interestingly, in many development and disease context, rare cell types often been see to play an important role although they only contribute to a small proportion of the cell population. For example, stem cells that contribute to new born neuron in adult brain are critical to reverse neurodegenerative diseases [1], and drug-resistant cells are the key barrier to cure cancer [2].

Genome-wide gene expression profiles now are widely accepted to define cell types. Recently technical advance of massive parallel single-cell RNA-seq on large-scale provides an unprecedented opportunity to discover previously unrecognized cell types due to their rarity. Although the number of cells being profiled single-cell transcriptome assay increase the chance that rare cell being sampled, tailored computational method to detect them remains highly demanded.

One of the major challenges is to identify genes that are associated with rare cell types without prior biological knowledge. GiniClust [3] adapted Gini index, an informative statistical measure widely used in social domain, to selecting rare cell-type-associated genes. It is implemented in Python and R and can be applied to

Guo-Cheng Yuan (ed.), *Computational Methods for Single-Cell Data Analysis*, Methods in Molecular Biology, vol. 1935, https://doi.org/10.1007/978-1-4939-9057-3_5, © Springer Science+Business Media, LLC, part of Springer Nature 2019

datasets originating from different platforms, such as multiplex qPCR data, traditional single-cell RNA-seq, or UMI-based single-cell RNA-seq, e.g., inDrops, Drop-seq, and 10× genomics.

## 2    Materials

### 2.1    Operating System

Operating system is preferred to be Linux or MAC. In windows, Cygwin is highly recommended to build a Linux-like environment. It can be downloaded from https://cygwin.com/install.html. For the installation of GiniClust and its dependencies, Internet access is necessary.

### 2.2    Python Packages

The graphical user interface of GiniClust relies on wxPython, a Python wrapper for the cross-platform wxWidgets API. Please ensure that you have Python 2.7 in your environment. In addition, GiniClust relies on the following libraries:

1. Gooey.
2. Setuptools.

Those packages should be automatically installed or upgraded via a pip installation. For instance, to install Gooey, proceed as follows:

3. Start a terminal session;
4. run $ pip install Gooey --upgrade.

If in doubt, please check that those libraries got installed properly by trying to import them or some of their modules in your Python interpreter: >>> import gooey, pkg_resources.

### 2.3    R Packages

As for the R code at the core of much of GiniClust's computations, for MAC and WINDOWS only the official R (3.2.1 or higher version) installation file is supported and tested. Using other installation methods, such as brew, may lead to running error. Besides, some users might experience issues installing another of GiniClust's dependencies: the MAST R package (*see* **Note 1**).

### 2.4    Input Files

The input file is a gene expression matrix in comma-separated value (csv) format. Specifically, for qPCR data, each row is log2 gene expression level; for RNAseq data, each row is UMI-Count/Cell or Raw-Read-Count/Cell (*see* **Note 2**). The first row contains cell IDs. The first column contains unique gene names. For example, user can take a look at one of our test datasets (stored in the sample_data folder within GiniClust's repository):

```
>ExprM.RawCounts<-read.csv("Data_GBM.csv", sep=",", head=T)
>ExprM.RawCounts[1:4,1:4]
#              MGH26 MGH26.1 MGH26.2 MGH26.3
#1/2-SBSRNA4 0      47       0        0
#A1BG        41     80       3        0
#A1BG-AS1    0      0        0        0
#A1CF        0      0        0        0
```

## 3  Methods

### 3.1  Run GiniClust Through Python-Based Graphical User Interface

To run GiniClust, please download the GiniClust GitHub repository from https://github.com/lanjiangboston/GiniClust/archive/master.zip, unzip it and move to the extracted directory so that it becomes your current working directory. Then, in a Linux environment, start a terminal session and enter:

```
$ python GiniClust.py.
```

From an OS X or Windows environment, launch a terminal session and enter:

```
$ pythonw GiniClust.py
```

A graphical user interface springs up and directs you into choosing a file to process from your arborescence of directories, specify the type of data by choose "qpcr" or "rna" under the "Actions" column, along with the name of the folder where you would like to store GiniClust's output (*see* the section below for more information about those files). A screenshot is provided (Fig. 1).

### 3.2  Run GiniClust Through R Script Main Function

Alternatively, GiniClust can be run directly as an R script at the command-line interface. User can run GiniClust in terminal session using Rscript like following:

```
$ Rscript Giniclust_Main.R [options]
```

You can specify the following options:

- -f CHARACTER or --file = CHARACTER, input dataset file name
- -t CHARACTER or --type = CHARACTER, input dataset type: choose from 'qPCR' or 'RNA-seq'
- -o CHARACTER or --out = CHARACTER, output folder name [default = results]

GiniClust

**Settings**
Detecting rare cell-types from single-cell gene expression data

**Actions**

qpcr

rna

**Required Arguments**

**Input**
Select a file to process:

/home/LanJiang/MyDatasets/Data_GBM.csv     Browse

**Optional Arguments**

**epsilon**
DBSCAN epsilon parameter:

0.5

**minPts**
DBSCAN minPts parameter:

3

**Output**
Specify GiniClust's output directory:

/home/LanJiang/MyGiniClustResults

Cancel     Start

**Fig. 1** Graphical user interface of GiniClust

- -e DOUBLE or --epsilon = DOUBLE, DBSCAN epsilon parameter qPCR:[default = 0.25],RNA-seq:[default = 0.5]
- -m INTEGER or --minPts = INTEGER, DBSCAN minPts parameter qPCR:[default = 5],RNA-seq:[default = 3]
- -h or --help, Show help message and exit.

For example, the following command is used to analyze the 'Data_GBM.csv' dataset.

```
$ Rscript GiniClust_Main.R -f Data_GBM.csv -t RNA-seq -o
GBM_results
```

The following command is used to analyze the 'Data_qPCR.csv' dataset.

```
$ Rscript GiniClust_Main.R -f Data_qPCR.csv -t qPCR -o
qPCR_results
```

Default parameters for GiniClust are shown as below:

```
minCellNum              = 3
minGeneNum              = 2000
expressed_cutoff        = 1
log2.expr.cutoffl       = 0
log2.expr.cutoffh       = 20
Gini.pvalue_cutoff      = 0.0001
Norm.Gini.cutoff        = 1
span                    = 0.9
outlier_remove          = 0.75
Gamma                   = 0.9
diff.cutoff             = 1
lr.p_value_cutoff       = 1e-5
CountsForNormalized     = 100000
rare_p                  = 0.05
perplexity              = 30
```

Usually default parameters work well for identifying rare cell types for datasets from different platforms. However, the user can always explore with different parameters. Users can use a text editor to open "yourworkdir/Rfunction/GiniClust_parameters.R." After changing parameters, users can rerun the GiniClust through Giniclust_Main.R script.

### 3.3  Run GiniClust Through R Script Step by Step

While GiniClust_Main.R is more friendly to use, running Giniclust step by step gives users more control and thus is more suitable for customized analysis. GiniClust stores intermediate files for each step. So it is convenient for users to check the results and to change parameters if necessary and rerun a single step before moving to the next one. A step-by-step instruction of running GiniClust is described in the following:

#### 3.3.1  Loading Packages

This step first loads all the additional required packages. If it is the first time running GiniClust, this step automatically installs all R packages, so it is necessary to have Internet access.

```
>source("Rfunction/GiniClust_packages.R")
```

Additionally, users can load all the R functions defined by GiniClust:

```
> source("Rfunction/GiniClust_Preprocess.R")
> source("Rfunction/GiniClust_Filtering.R")
> source("Rfunction/GiniClust_Fitting.R")
> source("Rfunction/GiniClust_Clustering.R")
> source("Rfunction/GiniClust_tSNE.R")
> source("Rfunction/DE_MAST.R")
> source("Rfunction/DE_t_test.R")
```

### 3.3.2  Preprocessing

This step is used for preprocessing and loading the input data. The input data for RNA-seq should be raw read counts or UMI counts. For qPCR data, since the input data are in the log2 scale, GiniClust transforms back to normal scale to consistently process the data for the later steps. The variables of input include data.type, out.folder, and exprimentID, while the file of input is "exprimentID_rawCounts.csv." And the variable of output is "ExprM.RawCounts," and is saved as a file named "exprimentID_rawCounts.csv."

```
> ExprM.RawCounts = GiniClust_Preprocess(data.file, data.type, out.folder,
  exprimentID)
```

### 3.3.3  Filtering the Preprocessed Data

GiniClust provides some basic filtering functions. The variable of input is "ExprM.RawCounts." The variable of output is "ExprM.RawCounts.filter." The intermediate stored file is "exprimentID_gene.expression.matrix.RawCounts.filtered.csv." The parameters involved and default values are:

```
> minCellNum        = 3
> minGeneNum        = 2000
> expressed_cutoff  = 1
```

After modifying the parameters (*see* **Note 3**), run the code below:

```
> ExprM.Results.filter=GiniClust_Filtering(ExprM.RawCounts, out.folder,
  exprimentID)
```

### 3.3.4  Selecting a Subset of the Genes for Clustering

This step normalizes the Gini index by LOESS curve fitting in Gini vs. Max space. The variable of input is "ExprM.RawCounts.filter." The variable of output is "Genelist.top_pvalue" or "Genelist.HighNormGini." And the intermediate stored files are "Gini_related_table_RNA-seq.csv" or "Gini_related_table_qPCR.csv." The parameters involved and default values are:

```
> log2.expr.cutoffl   = 0
> log2.expr.cutoffh   = 20
> Gini.pvalue_cutoff  = 0.0001
> Norm.Gini.cutoff    = 1
> span                = 0.9
> outlier_remove      = 0.75
```

After modifying the parameters (*see* **Note 4**), run the code below:

```
> GeneList.final = GiniClust_Fitting(data.type, ExprM.RawCounts.filter, out.folder,
  exprimentID)
```

*3.3.5   Clustering*

This step builds a cell-cell discard distance based on selected genes from **step 4**. Then DBSCAN is used to detect clusters. The variable of input is "ExprM.RawCounts.filter." The variable of output is "cell.cell.distance,", "c_membership," "clustering_membership_r," and "rare.cells.list.all." And the intermediate files are "exprimentID _clusterID.csv," "exprimentID _rare_cells_list.txt." The parameters involved and default values are:

```
> eps       = 0.5
> MinPts    = 3
> rare_p    = 0.05
```

After modifying the parameters (*see* **Note 5**), run the code below:

```
> Cluster.Results =GiniClust_Clustering(data.type, \
                ExprM.RawCounts.filter,\
                GeneList.final,eps,\
                MinPts, \
                out.folder,\
                exprimentID)
> cell.cell.distance      = Cluster.Results$cell_cell_dist
> c_membership            = Cluster.Results$c_membership
> clustering_membership_r = Cluster.Results$clustering_membership_r
> rare.cells.list.all     = Cluster.Results$rare.cell
```

Users can check the clustering results, for example,

```
> table(c_membership)
> print(rare.cells.list.all)
```

*3.3.6   tSNE Visualization*

A nonlinear dimension reduction technique called tSNE is used to visualize clustering results. The variable of input is "c_membership," "cell.cell.distance." The variable of output is "Rtnse_coord2." And the intermediate stored files are "exprimentID_Rtnse_coord2.csv." A figure is also generated and stored in the work folder (Fig. 2). The parameters involved and default values are:

```
> perplexity       = 30
```

After modifying the parameters (*see* **Note 6**), run the code below:

```
>GiniClust_tSNE(data.type, c_membership , cell.cell.distance, \
                perplexity,  out.folder, exprimentID)
```

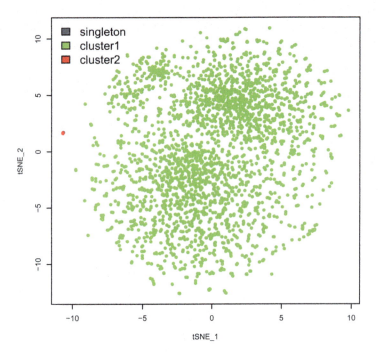

**Fig. 2** An example of tSNE visualization of clustering results

*3.3.7 Differentially Expressed Genes*

Comparing the single-cell expression data of each putative rare cell type cluster to the largest cluster will identify differentially expressed genes. The variable of input is "ExprM.RawCounts.filter," "rare.cells.list.all," "c_membership." The variable of output is "differential.r." And the intermediate stored files are "RareCluster.overlap_genes.txt" and "RareCluster_lrTest.csv" or "RareCluster.diff.gene.t-test.results.csv." The parameters involved and default values are:

```
diff.cutoff        = 1
lr.p_value_cutoff  = 1e-5
```

After modifying the parameters (*see* **Note 7**), for RNA-seq data, run the code below:

```
> DE_MAST(ExprM.RawCounts.filter, rare.cells.list.all, out.folder, exprimentID)
```

For qPCR data, run the code below:

```
> DE_t_test(ExprM.RawCounts.filter, rare.cells.list.all, c_membership, out.folder, exprimentID)
```

**3.4  Full List of Output Files and Description**

The output directory specified by the user at the graphical user interface contains the following files and directories.

Main results:

exprimentID_rare_cells_list.txt: the clusters of rare cells detected by Giniclust.

RareCluster_lrTest.csv or RareCluster.diff.gene.t-test.results. csv: Differentially expressed genes results for the rare cells type cluster.

Other supporting results:

exprimentID_rawCounts.csv: the raw counts.

exprimentID_normCounts.csv: the normalized counts.

exprimentID_gene.expression.matrix.RawCounts.filtered.csv: the raw counts after filtering.

exprimentID_gene.expression.matrix.normCounts.filtered. csv: the normalized counts after filtering.

Gini_related_table_RNA-seq.csv: the table related with Gini index for RNA-seq data.

Gini_related_table_qPCR.csv: the table related with Gini index for qPCR data.

exprimentID_clusterID.csv: clustering result, the first column represents cell IDs and the second column is the corresponding cluster result for each cell.

exprimentID_Rtnse_coord2.csv: coordinates of cells in tSNE plot.

exprimentID_bi-directional.GiniIndexTable.csv: For qPCR data the table of bidirectional Gini index.

RareCluster.overlap_genes.txt: overlap genes between the selected high Gini genes and DE genes in rare cluster.

Sub-folder 'figures':

exprimentID_histogram of Normalized.Gini.Socre.pdf: histogram of estimated $p$-values based on a normal distribution approximation for genes.

exprimentID_smoothScatter_pvalue_gene.pdf: the smoothScatter plot in which the red points are the selected high Gini genes according to specified cutoff.

exprimentID_tsne_plot.pdf: tSNE plot for cells.

exprimentID_RareCluster_diff_gene_overlap.pdf: Venn diagram for differentially expressed genes and high gini genes.

exprimentID_RareCluster_overlapgene_rawCounts_bar_plot. genename.pdf: barplot of rare cluster and major cluster for the overlap genes.

**3.5  Further Reading**

It is worth noting that while the GiniClust is powerful for detecting rare cell types cluster, it is not sensitive for distinguishing major cell types. This limitation can be partially resolved by updated version of GiniClust called GiniClust2. It uses a novel cluster-aware weighted consensus clustering algorithm to combine GiniClust and Fano-based k-means clustering results, by maximizing the strengths of these individual clustering methods in detecting rare and common clusters, respectively [4].

## 4   Notes

1. If users have a problem with installing MAST packages, please visit the website (https://github.com/RGLab/MAST) for detailed instructions. We recommend that users upgrade MAST package to the newest version. If you are using an old version, you may need to replace the file DE_MAST.R by https://github.com/lanjiangboston/GiniClust/blob/master/Archive/DE_MAST.R.

2. log2 transformed RNA-seq data for Giniclust may not work. We suggest that users use featureCounts from http://subread.sourceforge.net/ [5] or htseq-count from http://www-huber.embl.de/users/anders/HTSeq/doc/counting.html [6] to get raw reads counts.

3. minCellNum means the minimum number of cells for rare cell clusters. It is highly recommended that this value is set to be equal to or larger than 3. However, the larger value of this parameter means less sensitivity. minGeneNum is used for filtering cell that may not express enough genes. The default value of 2000 is consistent with recent report about scRNA-seq using smFISH as guide [7].

4. "log2.expr.cutoffl' " and "log2.expr.cutoffh" define the range of gene expression. "Gini.pvalue_cutoff" controls how many genes finally are chosen. "Norm.Gini.cutoff" controls choosing high gini genes based on $p$-value or not (default $= 1$). "Span" and "outlier_remove" are parameters used in LOESS fitting.

5. "eps" and "MinPts" are parameter of "epsilon" and "minimum points" for DBSCAN, respectively. The distance of any point to its nearest core point of the same cluster is less than "epsilon." larger values for "MinPts" are usually better for data sets with noise and form more significant clusters, and however, lose the sensitivity for rare cell clusters. "rare_p" is a parameter to define what you call rare cell cluster. For example, rare_$p = 0.05$ means only cluster that contributes to less than 5% of the whole population is called rare cell cluster.

6. "Perplexity" is a parameter for tSNE. The results of t-SNE are fairly robust for different perplexity. The most appropriate value depends on the density and size of the data. Typical values for the perplexity range between 5 and 50.

7. "diff.cutoff" and "lr.p_value_cutoff" are used to filter differential expressed genes based on log2 fold change and $p$ value during MAST analysis, respectively. Increasing the value of "diff.cutoff" or decreasing the value of "lr.p_value_cutoff" will decrease the number of differential expressed genes.

# References

1. Habib N, Li Y, Heidenreich M, Swiech L, Avraham-Davidi I, Trombetta JJ, Hession C, Zhang F, Regev A (2016) Div-Seq: single-nucleus RNA-Seq reveals dynamics of rare adult newborn neurons. Science 353 (6302):925–928. https://doi.org/10.1126/science.aad7038

2. Sharma SV, Lee DY, Li B, Quinlan MP, Takahashi F, Maheswaran S, McDermott U, Azizian N, Zou L, Fischbach MA, Wong KK, Brandstetter K, Wittner B, Ramaswamy S, Classon M, Settleman J (2010) A chromatin-mediated reversible drug-tolerant state in cancer cell subpopulations. Cell 141 (1):69–80. https://doi.org/10.1016/j.cell.2010.02.027

3. Jiang L, Chen H, Pinello L, Yuan GC (2016) GiniClust: detecting rare cell types from single-cell gene expression data with Gini index. Genome Biol 17(1):144. https://doi.org/10.1186/s13059-016-1010-4

4. Tsoucas D, Yuan GC (2018) GiniClust2: a cluster-aware, weighted ensemble clustering method for cell-type detection. Genome Biol 19(1):58. https://doi.org/10.1186/s13059-018-1431-3

5. Liao Y, Smyth GK, Shi W (2014) Featurecounts: an efficient general purpose program for assigning sequence reads to genomic features. Bioinformatics 30(7):923–930. https://doi.org/10.1093/bioinformatics/btt656

6. Anders S, Pyl PT, Huber W (2015) HTSeq--a Python framework to work with high-throughput sequencing data. Bioinformatics 31 (2):166–169. https://doi.org/10.1093/bioinformatics/btu638

7. Torre E, Dueck H, Shaffer S, Gospocic J, Gupte R, Bonasio R, Kim J, Murray J, Raj A (2018) Rare cell detection by single-cell RNA sequencing as guided by single-molecule RNA FISH. Cell Syst 6(2):171–179.e175. https://doi.org/10.1016/j.cels.2018.01.014

# Chapter 6

# scMCA: A Tool to Define Mouse Cell Types Based on Single-Cell Digital Expression

## Huiyu Sun, Yincong Zhou, Lijiang Fei, Haide Chen, and Guoji Guo

## Abstract

For decades, people have been trying to define cell type with the combination of expressed genes. The choice of the limited number of genes for the classification limits the precision of this system. Here, we build a "single-cell Mouse Cell Atlas (scMCA) analysis" pipeline based on scRNA-seq datasets covering all mouse cell types. We build the scMCA reference and then use the tool "scMCA" to match single-cell digital expression to its closest cell type.

**Key words** scMCA, Mouse Cell Atlas, scRNA-seq

## 1 Introduction

Single-cell RNA sequencing (scRNA-seq) is a powerful tool to perform transcriptomes analyses at the single-cell level. It can reveal the gene expression status of individual cells and capture rare populations that are difficult to obtain with conventional bulk RNA-seq data. Recently, Han and colleagues [1] have performed scRNA-seq on 400 k single cells from 51 mouse tissues, organs, and cell cultures. The study provides an initial draft of the mouse cell atlas (MCA) with a comprehensive database that contains transcriptional characteristics of almost all major cell types in mouse. It is therefore possible to construct a transcriptomic reference to map and define unknown cell types.

Here, we established a database of more than 800 mouse cell types using scRNA-seq data from MCA [1] together with other studies [2–6]. After selecting differentially expressed genes and grouping the cell type clusters, we use averaged expression values of the same group to construct a cell-type reference. A pipeline

Huiyu Sun, Yincong Zhou, Lijiang Fei, and Haide Chen contributed equally to this work.

Guo-Cheng Yuan (ed.), *Computational Methods for Single-Cell Data Analysis*, Methods in Molecular Biology, vol. 1935, https://doi.org/10.1007/978-1-4939-9057-3_6, © Springer Science+Business Media, LLC, part of Springer Nature 2019

called scMCA analysis was established to identify the cell types at single-cell level. The pipeline can also be extended to identify bulk RNA-seq data.

## 2  Materials

### 2.1  scRNA-Seq Datasets Resource

We use scRNA-seq data with more than 800 mouse cell clusters from different tissues to build our scMCA reference (*see* Table 1). The datasets we used to build the reference are listed as below, details of datasets are available via MCA website http://bis.zju.edu.cn/MCA/gallery.html.

### 2.2  Software and Algorithms

You can use MCA website or R package scMCA to define the mouse cell types of your data. The scMCA on MCA website is available at http://bis.zju.edu.cn/MCA/blast.html. We have tested it on Firefox, Chrome, and Safari. For huge digital gene expression (DGE) data (the DGE file larger than 200 MB), the R package is available on our Github page: https://github.com/ggjlab/scMCA. The softwares and algorithms used for scMCA are listed in Table 2.

**Table 1**
**The datasets used for building scMCA reference**

| Datasets | Reference |
|---|---|
| Datasets of microwell-seq data (more than 40 tissues) | Han et al. [1] |
| Dataset of Arc-ME (the complex of the arcuate hypothalamus and median eminence) | Campbell et al. [2] |
| Dataset of pancreatic islets | Baron et al. [3] |
| Dataset of lung mesenchyme | Zepp et al. [4] |
| Dataset of whole mouse E8.25 embryos | Ibarra-Soria et al. [5] |
| Dataset of retina | Macosko et al. 2015 [6] |

**Table 2**
**The packages used for scMCA**

| Packages | Source |
|---|---|
| Pheatmap | https://cran.r-project.org/web/packages/pheatmap/index.html |
| Shiny | http://shiny.rstudio.com/ |
| Dplyr | https://cran.r-project.org/web/packages/dplyr/index.html |

## 3 Methods

### 3.1 Establish the Reference with Different scRNA-Seq Datasets (See Note 1)

#### 3.1.1 Normalization of Different scRNA-Seq Datasets

Before downstream analyses, the normalization is needed. The DGE is processed by the following formula:

$$E = Count*100,000/sum\ (Count).$$

$E$: the normalizing expression values;

Count: the raw UMI counts;

sum (Count): to sum up all counts in one cell.

#### 3.1.2 Define Reference Values of Cell Clusters

By default, we randomly choose 100 cells (all cells will be chosen if the total number is less than 100) from each cell cluster for 3 times to get the gene average expression values (*see* **Note 2**), and then take the integer of the values for each sample (*see* **Note 3**). The average of 3 representative values is used as cell cluster value in reference.

#### 3.1.3 Feature Gene Selection

We used the integer values in three representative averaged data to perform differential gene expression analysis. Top 10 genes (avg_logFC > 1) of each cell cluster were merged to make the reference feature gene list (3028 feature genes). The differential gene expression analysis was done by the Wilcoxon rank sum test method using R Package Seurat.

#### 3.1.4 Get the MCA Cell Type Reference

The representative values of more than 800 cell clusters with 3028 features genes were used as MCA cell type reference, and the reference was log-transformed before calculating scMCA scores.

### 3.2 Using scMCA on MCA Website

#### 3.2.1 Submit RNA-Seq DGE File to scMCA Website

After uploading the scRNA-seq or bulk RNA-seq DGE file, you should click the "scMCA" button to perform scMCA pipeline (Fig. 1). The uploaded file should be a DGE matrix in.txt or.csv format with each row representing a gene, and each column representing a cell (or sample). The numerical expression can be raw count, FPKM, and RPKM. After submission, the DGE matrix will be log-normalized (*see* **Note 4**) and the feature gene expression matrix will be extracted (*see* **Note 5**).

#### 3.2.2 Get Defined Results from scMCA Website

Pearson correlation coefficients of the extracted expression matrix against MCA cell type reference are calculated, and the correlation coefficients are used as the scMCA scores in scMCA pipeline. The scMCA results can be shown by interactive heatmap, in which the row means the defined cell type, the column means the query cell, and the color of block indicates the strength of the correlation (Fig. 2). One can also download the result table in a csv format,

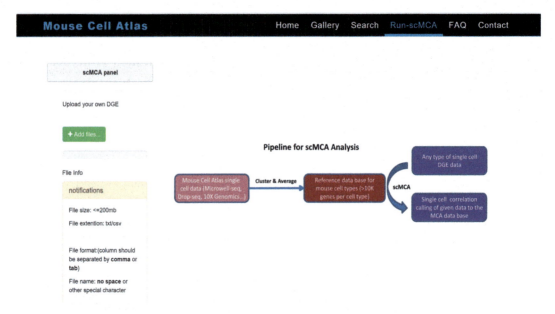

**Fig. 1** Data submission interface of scMCA on MCA website. You can upload DGE file in txt or csv format by clicking the "Add files" button and perform scMCA pipeline by clicking the "scMCA" button

**Fig. 2** The scMCA results of website. The result can be shown by interactive heatmap, in which the row means the defined cell type, the column means the query cell, and the colors of blocks indicate the strength of the correlations. You can also click the "result from scMCA" button to download the file which records the scMCA scores between query cells and cell types. The uploaded example dataset of distal lung epithelium is from Treutlein et al. [7]

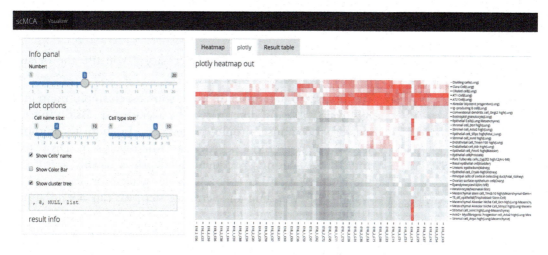

**Fig. 3** The scMCA results of R package. You can choose option buttons on the left of web interface to adjust the results. The uploaded example dataset of distal lung epithelium is from Treutlein et al. [7]

which records scMCA scores between query cells (cells to be identified) and reference cell types.

### 3.3  Using R Package scMCA

The scMCA package is hosted on Github. It can be conveniently installed via "devtools" by typing "devtools::install_github ("ggjlab/scMCA")."

*3.3.1  Installation of R Package scMCA*

*3.3.2  Usage of R Package scMCA*

scMCA package has two main functions, scMCA and scMCA_vis. scMCA calculates the Pearson correlation coefficient between each query cell and cell type. scMCA_vis is used to visualize the result returned from scMCA. To use scMCA, you should take the following steps:

1. Loading the scRNA-seq or bulk RNA-Seq DGE file to R environment.

2. Setting the number of most relevant cell types for each query cell. The corresponding parameter is "number_plot" in function scMCA. You can type "?scMCA" in R for more information.

3. Execute the scMCA and use scMCA_vis to get the results. Using scMCA_vis, you can open a web page in localhost which reflects the results of scMCA (Fig. 3).

For more details, you can find the instructions of R package scMCA on Github page.

## 4    Notes

1. The normalization step is unnecessary if you use the default mouse cell atlas reference.

2. We integrated data from different sources, clustered them into 894 cell types, and determined the average expression in each cluster for transcriptome references. In high-throughput scRNA-seq experiments, such as Microwell-seq, Drop-seq, and 10 × Genomics, sequencing depth is usually sacrificed; the average number of detectable genes for each cell is ~1000. After adding an averaging step, we can obtain about 10,000 genes in one cluster.

3. The number of detected genes in different cell types is different. To reduce such variation, we take the integer of average expression values from 100 randomly sampled cells for each cell type.

4. Log-normalization of RNA-seq DGE matrix is obtained by the following formula:

   $E = \log(Count*100,000/\text{sum}(Count) + 1)$.
   $E$: the log-normalized expression values;
   Count: the raw UMI counts (FPKM or RPKM);
   sum (Count): to sum up all counts in one cell.

5. Get the feature gene expression matrix of RNA-seq data. The expression matrix of 3028 feature genes was extracted from the log-normalized RNA-seq expression matrix. If the expression of a characteristic gene is not detected in the submitted data, the expression value of this characteristic gene is considered to be zero.

## References

1. Han X, Wang R, Zhou Y et al (2018) Mapping the mouse cell atlas by microwell-Seq. Cell 172:1091–107.e17

2. Campbell JN, Macosko EZ, Fenselau H et al (2017) A molecular census of arcuate hypothalamus and median eminence cell types. Nat Neurosci 20:484–496

3. Baron M, Veres A, Wolock SL et al (2016) A single-cell transcriptomic map of the human and mouse pancreas reveals inter- and intra-cell population structure. Cell Syst 3:346–60.e4

4. Zepp JA, Zacharias WJ, Frank DB et al (2017) Distinct Mesenchymal lineages and niches promote epithelial self-renewal and myofibrogenesis in the lung. Cell 170:1134–48.e10

5. Ibarra-Soria X, Jawaid W, Pijuan-Sala B et al (2018) Defining murine organogenesis at single-cell resolution reveals a role for the leukotriene pathway in regulating blood progenitor formation. Nat Cell Biol 20:127–134

6. Macosko EZ, Basu A, Satija R et al (2015) Highly parallel genome-wide expression profiling of individual cells using nanoliter droplets. Cell 161:1202–1214

7. Treutlein B, Brownfield DG, Wu AR et al (2014) Reconstructing lineage hierarchies of the distal lung epithelium using single-cell RNA-seq. Nature 509:371–375

# Chapter 7

## Differential Pathway Analysis

### Jean Fan

### Abstract

Integrating prior knowledge of pathway-level information can enhance power and facilitate interpretation of gene expression data analyses. Here, we provide a practical demonstration of the value of gene set or pathway enrichment testing and extend such techniques to identify and characterize transcriptional subpopulations from single-cell RNA-sequencing data using pathway and gene set overdispersion analysis (PAGODA).

**Key words** Single cell, Pathway, Gene set enrichment analysis, Differential expression analysis, Clustering

## 1 Introduction

Identifying genes that exhibit significant differences among two or more biological states, conditions, or cell-types is integral to understanding the putative molecular bases for phenotypic variation. Determining whether individual genes exhibit significant expression differences between conditions can be achieved using differential gene expression analysis [1]. However, when gene expression data are noisy and biological signals are weak, testing individual genes for differences may not provide any statistically significant results. In particular for single-cell RNA-seq data, such a differential expression analysis is often complicated by high levels of technical noise and intrinsic biological stochasticity in the data. As such, application of previous differential expression analysis approaches developed for bulk RNA-seq may not always be suitable [2]. While methods for differential expression analysis specifically tailored to single-cell RNA-seq data have been developed [3, 4], alternatively grouping genes into biologically-relevant modules such as pathways may greatly enhance statistical power and improve our ability to identify true biological signal [5, 6]. In this chapter, we will discuss how to take a pathway-informed approach to differential

Guo-Cheng Yuan (ed.), *Computational Methods for Single-Cell Data Analysis*, Methods in Molecular Biology, vol. 1935, https://doi.org/10.1007/978-1-4939-9057-3_7, © Springer Science+Business Media, LLC, part of Springer Nature 2019

expression analysis and apply it to single-cell RNA-seq data to identify and characterize transcriptional subpopulations.

Gene set or pathway enrichment analysis is a computational approach that determines whether an a priori defined set of genes such as a pathway shows statistically significant, concordant differences between two biological states. Gene set or pathway enrichment analysis is particularly powerful when genes individually do not exhibit a statistically significant difference between two biological states, but, when grouped together, show statistically significant concordant differences. For example, when performing differential expression analysis, one common approach is to use a significance cutoff to identify a limited number of the most interesting genes for further research and interpretation. Gene set or pathway enrichment analysis takes an alternative approach by focusing on cumulative expression changes of multiple genes as a group, thus shifting the focus from individual genes to groups of genes. By looking at several genes at once, such an approach can identify gene sets or pathways that have several genes each change a small amount, but in a coordinated way, which may reach statistical significance even when individual gene expression changes are quite small and insufficiently significant.

There are many different methods to perform gene set or pathway enrichment analysis, from hypergeometric distribution tests [7, 8] to permutation-based approaches [5]. One popular method, aptly named Gene Set Enrichment Analysis (GSEA) [5], tests for enrichment using a permutation-based approach. In GSEA, first, genes are ranked such as based on a measure of each gene's differential expression with respect to the two conditions. Then the entire ranked list is used to assess how the genes of each gene set are distributed across the ranked list by walking down the ranked list of genes, increasing a running-sum statistic when a gene belongs to the set, and decreasing it when the gene does not (Fig. 1). The enrichment score is the maximum deviation from zero encountered during the walk. The score reflects the degree to which the genes in a gene set are overrepresented at the top or bottom of the entire ranked list of genes. A set that is not enriched will have its genes spread more or less uniformly through the ranked list. An enriched set, on the other hand, will have a larger portion of its genes at one or the other end of the ranked list. The extent of enrichment is captured mathematically as the score statistic. The statistical significance of the score can then be estimated using permutation, whereby enrichment scores are computed for random gene sets of the same size as the tested gene set. This randomization is repeated many times to produce an empirical null distribution of scores. The nominal $p$-value estimates the statistical significance of a single gene set's score based on the permutation-generated null distribution.

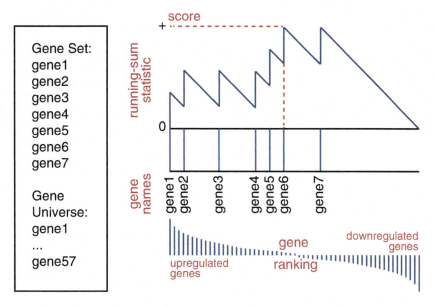

**Fig. 1** A standard gene set enrichment plot. Genes in the gene universe are ranked according to a differential expression statistic from most upregulated to most downregulated. A running-sum statistic then traverses the ranked list and increments the enrichment score statistic upon reaching a gene within the gene set of interest

## 2 Materials

All programming will be done using the R statistical programming language [9].

**2.1 Liger R Package**

In Subheading 3.1, we will perform gene set enrichment analysis using the Lightweight Iterative Gene set Enrichment in R (liger) package, an R implementation of the GSEA algorithm [5]. liger can be installed from CRAN using the following command in R:

```
install.packages("liger")
```

**2.2 Scde R Package**

In Subheadings 3.2 and 3.3, we will perform pathway and gene set overdispersion analysis (PAGODA) using the Single Cell Differential Expression (scde) package. Scde can be installed from Bioconductor using the following command in R:

```
# try http:// if https:// URLs are not supported
source("https://bioconductor.org/biocLite.R")
biocLite("scde")
```

## 3    Methods

### 3.1    Enhancing Statistical Power by Incorporating Pathway-Level Information

To demonstrate the utility of gene set or pathway enrichment analysis, we will use a simulated dataset. Specifically, we will simulate a weak differential expression within a known gene set between two biological samples. We will show that while differential expression analysis is not able to pick up these genes as significantly differentially expressed, a gene set enrichment analysis will be able to pick up significant enrichment.

1. First, we will load the liger package.

```
library(liger)
```

2. Load a gene set based on Gene Ontology (GO) terms.

```
# load gene set
data("org.Hs.GO2Symbol.list")
```

We can look into the newly loaded org.Hs.GO2Symbol.list object. Notice that it is a list of GO ids for various gene sets. Each list contains the human HUGO symbols of genes within that gene set. Note, in this manner, alternative gene sets such as MSigDB [10] or KEGG or even custom gene sets can also be created and used.

```
head(org.Hs.GO2Symbol.list)
## $`GO:0000002`
##  [1] "AKT3"     "C10orf2"  "DNA2"     "LIG3"     "MEF2A"    "MGME1"
##  [7] "MPV17"    "OPA1"     "PID1"     "PRIMPOL"  "SLC25A33" "SLC25A36"
## [13] "SLC25A4"  "STOML2"   "TYMP"
...
```

3. To simulate a weak differential expression within a known gene set between two biological samples, we will first simulate random gene expression for 100 cells. We will create a matrix containing all genes and simulate gene expression by drawing from a normal distribution with mean = 0, and sd = 3. Alternatively, your own normalized single-cell expression data can be substituted in at this step.

```
# set random seed to ensure reproducibility
set.seed(0)
# get universe of genes
universe <- unique(unlist(org.Hs.GO2Symbol.list))
# make random data
Nsamples <- 100 # 100 cells
Mgenes <- length(universe)
mat <- matrix(rnorm(Mgenes*Nsamples, mean=0, sd=3), Mgenes, Nsamples)
rownames(mat) <- universe
```

Next, we will first pick a gene set.

```
# get genes in gene set GO:0000002
gs <- org.Hs.GO2Symbol.list[["GO:0000002"]]
# genes
print(gs)
##  [1] "AKT3"     "C10orf2"  "DNA2"     "LIG3"     "MEF2A"    "MGME1"
##  [7] "MPV17"    "OPA1"     "PID1"     "PRIMPOL"  "SLC25A33" "SLC25A36"
## [13] "SLC25A4"  "STOML2"   "TYMP"
```

We will split our 100 cells into 2 groups to represent two biologically different states.

```
# two biological states (groups)
group <- factor(c(rep(1, Nsamples/2), rep(2, Nsamples/2)))
names(group) <- colnames(mat) <- paste0('sample', 1:Nsamples)
```

Now, we will simulate upregulation of genes from our selected gene set in cells belonging to group 1. Genes from our selected gene set in cells belonging to group 1, rather than being drawn from a normal distribution with mean = 0 and sd = 3, will instead have an increased mean = 2.25, and sd = 5. We will also remove negative values in our simulation to keep simulated expression values interpretable.

```
# simulate upregulation of gene set in group 1
mat[gs, group==1] <- rnorm(length(gs)*sum(group==1), mean=2.25, sd=5)
# make more realistic; can't have negative gene expression
mat[mat < 0] <- 0
```

4. Now we can visualize the expression of our simulated upregu-lated genes along with 50 other non-upregulated genes using a heatmap (Fig. 2). We will also color the column side bar of our heatmap using the cell group labels, with group 1 cells labeled in red, and group 2 cells labeled in blue. Similarly, we will color the row side bar in green if the gene is within our selected gene set and black if not.

```
# we can visualize this weak differential expression in a heatmap
# visualize weakly differentially expressed genes and another 50 genes
vi <- c(gs, universe[1:50])
# label supposedly differentially expressed genes
heatmap(mat[vi,], Rowv=NA, Colv=NA, scale="none",
        col=colorRampPalette(c("blue", "white", "red"))(100),
        RowSideColors = c('black', 'green')[as.factor(vi %in% gs)],
        ColSideColors = c('red', 'blue')[group])
```

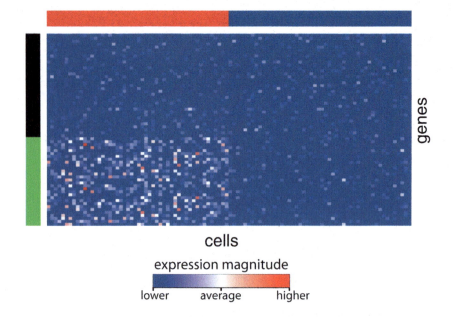

**Fig. 2** Gene expression heatmap for select simulated genes. Rows are genes and columns are cells. Gene expression is colored using a color ramp from blue to white to red, with highly expressed genes colored in red and lowly expressed genes in blue. Column side bar is colored using the cell group labels, with group 1 cells labeled in red, and group 2 cells labeled in blue. Row side bar is colored in green if the gene is within our selected gene set and black if not

Although we simulated the green row side color annotated genes to be upregulated in the red column side color annotated samples compared to the blue column side color annotated samples, even visually, it is somewhat difficult to tell which genes are differentially expressed.

5. We can also quantify the extent of the differential expression between our two biological states using a t-test.

```r
# run differential expression analysis using simple t-test
vals.info <- lapply(1:nrow(mat), function(i) {
    pv <- t.test(
        mat[i, group==1],
        mat[i, group==2]
    )
    return(list(val=pv$statistic, p=pv$p.value))
})
vals <- unlist(lapply(vals.info, function(x) x$val))
p <- unlist(lapply(vals.info, function(x) x$p))
names(p) <- names(vals) <- rownames(mat)
```

6. Because we are testing many genes, we have to apply multiple-testing correction. We will use a Bonferroni correction [11].

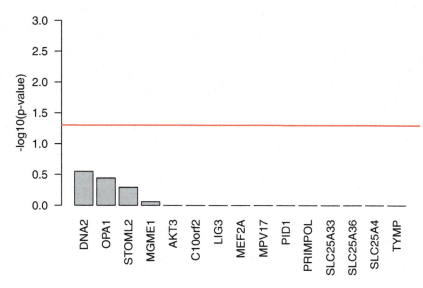

**Fig. 3** Differential expression analysis results for select simulated genes. Barplot shows -log10( $p$ -value) for each gene. Red line shows the $p = 0.05$ significance threshold. Note none of the tested genes passes the significance threshold

```
p.adj <- p.adjust(p, method="bonferroni") # multiple-testing correction
names(p.adj) <- rownames(mat)
```

7. We can now visualize the final -log10( $p$ -values) using a barplot (Fig. 3). We will use a red line to indicate the common $p < 0.05$ significance threshold. Significant genes should have bars that pass the red line.

```
barplot(sort(-log10(p.adj[gs]), decreasing=TRUE), ylim=c(0, 3), las=2)
abline(h = -log10(0.05), col="red")
```

Unfortunately, none of the genes, including those we simulated to be differentially expressed, were actually picked up as significantly differentially expressed after multiple-testing correction (with corrected $p$ -values <0.05). In a real-world situation, we may be tempted to end our analysis here and conclude that since nothing is significantly differentially expressed between the two biological states there is no significant difference.

However, we can still perform gene set or pathway enrichment analysis on a priori defined gene sets to look for statistically significant concordant differences.

8. We will perform such analyses using liger for 10 Gene Ontology gene sets in org.Hs.GO2Symbol.list, including GO:0000002.

```
# run iterative bulk gsea on our true gene set and 9 other gene sets as
test
gseaVals <- iterative.bulk.gsea(
    values = vals,
    set.list = org.Hs.GO2Symbol.list[1:10],
    rank=TRUE)
## initial: [1e+02 - 3] [1e+03 - 1] [1e+04 - 1] done
print(gseaVals)
##                   p.val       q.val      sscore        edge
## GO:0000002 0.00009999 0.00059994   2.5584741   2.0848842
## GO:0000003 0.66336634 0.66336634   0.4924230   0.3374948
## GO:0000012 0.11888112 0.25774226  -0.9737758  -0.1256842
## GO:0000014 0.24752475 0.36831683   0.6518057  -0.5915193
## GO:0000018 0.30693069 0.36831683  -0.7279604   0.9366813
## GO:0000022 0.12887113 0.25774226   0.9455950  -0.4223886
```

9. We can then identify significantly enriched gene sets as those with a $q$-value $<0.05$.

```
# identify significantly enriched gene sets
gseaSig <- rownames(gseaVals[gseaVals$q.val < 0.05,])
print(gseaSig)
## [1] "GO:0000002"
```

Indeed, we recover GO:0000002 as a significantly enriched gene set!

10. We can visualize a standard gene set enrichment plot for this gene set (Fig. 4).

```
# look at plots
for(i in seq_along(gseaSig)) {
  gs <- org.Hs.GO2Symbol.list[[gseaSig[i]]]
  gsea(values=vals, geneset=gs, mc.cores=1, plot=TRUE, rank=TRUE)
}
```

So, although no individual gene was found to be statistically significantly differentially expressed between our two biological states, gene set and pathway enrichment analysis identified a significantly enriched gene set, GO:0000002, which is exactly the gene set that we simulated to show concordant differences. By looking for coordinated changes in genes within these a priori defined gene sets, we are able to increase our statistical power to identify differences between our two biological states.

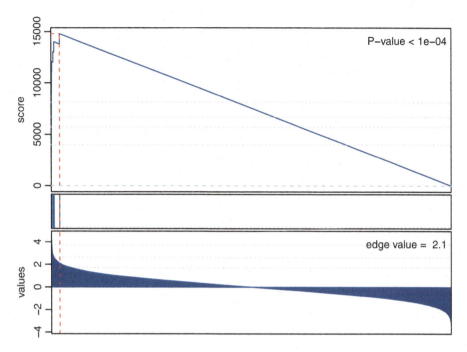

**Fig. 4** Gene set enrichment plot for gene set GO:0000002 demonstrates significant enrichment as simulated

### 3.2 Applying a Pathway-Integrated Approach with Pathway and Gene Set Overdisperrsion Analysis

Gene set testing with methods such as liger can be used for differential expression analysis to increase statistical power and uncover likely functional interpretations. However, such testing requires knowledge of biological conditions or subpopulations for comparison. To identify these transcriptionally distinct subpopulations, a similar rationale can be applied in single-cell RNA-seq data analysis. Highly variable genes may partition cells into transcriptionally distinct subpopulations but carry consideration uncertainty as observed variability in gene expression may be the result of technical artifacts such as drop-outs. Yet whereas variability in the expression of a single gene may be noisy, coordinated upregulation of many genes within a gene set or pathway in the same subset of cells could provide a prominent signature to distinguish subpopulations.

Pathway And Gene set Over-Dispersion Analysis (PAGODA) [6] looks for coordinated expression variability of genes in both annotated pathways and automatically detected "de novo" gene sets. PAGODA then uses this gene set and pathway-level information to cluster cells into transcriptional subpopulations.

Briefly, PAGODA first estimates the effective sequencing depth, drop-out rate, and amplification noise of each cell using a previously described mixture-model approach with minor enhancements. Using these models, the observed expression variance of each gene is renormalized on the basis of the expected genome-wide variance at the appropriate expression magnitude. PAGODA then examines an extensive panel of gene sets to identify those showing a statistically significant excess of coordinated variability. Gene sets can include annotated pathways, such as Gene Ontology (GO) categories, as well as clusters of transcriptionally correlated genes found in a given data set ("de novo" gene sets). The prevalent transcriptional signature of each gene set is captured by its first principal component (PC), with weighted PCA used to adjust for technical noise. If the amount of variance explained by the first PC of a given gene set is significantly higher than expected, the gene set is considered to be "overdispersed. "PCs from the resulting significantly overdispersed gene sets are combined to form a single "aspect" of heterogeneity to provide a nonredundant view of transcriptional heterogeneity to users through an interactive web browser interface.

To demonstrate the utility of PAGODA, we will continue our exploration of our simulated dataset. Note, to run PAGODA using your own single-cell RNA-seq data, see Subheading 3.3 for step-by-step instructions on how to go from gene expression counts to the appropriate variance-normalized gene expression matrix inputted into the pathway overdispersion testing step.

1. We will first show how unbiased hierarchical clustering on our simulated raw data fails to cluster our true groups together (Fig. 5).

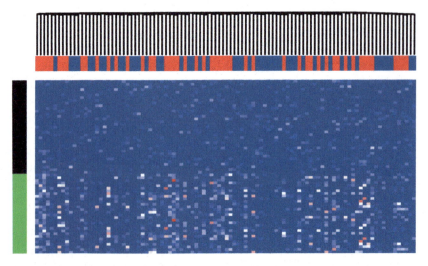

**Fig. 5** Gene expression heatmap with cells grouped by hierarchical clustering shows inconsistency with cell group labels. Rows are genes and columns are cells. Gene expression is colored using a color ramp from blue to white to red, with highly expressed genes colored in red and lowly expressed genes in blue. Column side bar is colored using the cell group labels, with group 1 cells labeled in red, and group 2 cells labeled in blue. Row side bar is colored in green if the gene is within our selected gene set and black if not

```
# just cluster by all genes
hc <- hclust(dist(t(mat)))
heatmap(mat[vi,], Rowv=NA, Colv=as.dendrogram(hc), scale="none",
        col=colorRampPalette(c("blue", "white", "red"))(100),
        RowSideColors = c('black', 'green')[as.factor(vi %in% gs)],
        ColSideColors = c('red', 'blue')[group])
```

The cells are ordered by unbiased hierarchical clustering and we do not see any segregation of our two cell group labels. However, we can integrate pathway-level information to enhance our signal and enable proper separation of our two simulated cell groups.

2. To run PAGODA, we will load the scde package and format our simulated data into the appropriate format. Note, to run PAGODA using your own single-cell RNA-seq data, additional functions are available for error-modeling and normalization from gene expression counts (*See.*

```
library(scde)

# format data to pipe into PAGODA
varinfo <- list()
varinfo$mat <- mat
matw <- matrix(1, nrow(mat), ncol(mat)) # equal weighting
rownames(matw) <- rownames(mat)
colnames(matw) <- colnames(mat)
varinfo$matw <- matw
```

3. We will compute PCs for a set of 10 pathways and cluster based on such pathway-level expression.

```
go.env <- list2env(org.Hs.GO2Symbol.list[1:10]) # just use first 10
pathways
# test pathways for overdispersion
pwpca <- pagoda.pathway.wPCA(varinfo, go.env, n.components = 1, n.cores =
1)
df <- pagoda.top.aspects(pwpca, return.table = TRUE, plot = FALSE,
z.score = 1.96)
head(df)
##          name npc   n     score          z     adj.z sh.z adj.sh.z
## 1 GO:0000002   1  15 212.82838 152.43638 152.42917   NA       NA
## 3 GO:0000018   1  34  31.30615  50.76668  50.75870   NA       NA
## 2 GO:0000003   1 300  10.29349  50.41152  50.41152   NA       NA

tam <- pagoda.top.aspects(pwpca, z.score = qnorm(0.01/2, lower.tail =
FALSE))
```

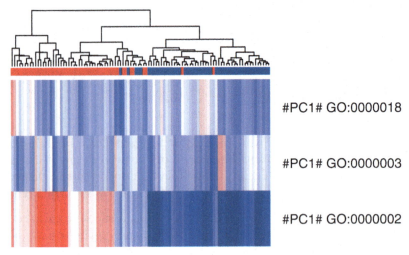

**Fig. 6** Pathway expression heatmap with cells grouped by hierarchical clustering shows consistency with cell group labels. Rows are pathways and columns are cells. Pathway expression, summarized by the first principal component (PC1) of gene expressions for genes within the pathway, is colored using a color ramp from blue to white to red. Column side bar is colored using the cell group labels, with group 1 cells labeled in red, and group 2 cells labeled in blue

4. Now, we can cluster our cells based on their pathway-level expression patterns using hierarchical clustering and visualize the results as a heatmap (Fig. 6).

```
# unbiased clustering on pathway information
hc2 <- hclust(dist(t(tam$xv)))
heatmap(tam$xv, Rowv=NA, Colv=as.dendrogram(hc2), scale="none",
        col=colorRampPalette(c("blue", "white", "red"))(100),
        ColSideColors = c('red', 'blue')[group], mar=c(5,15))
```

And indeed, we can see that the pathway-integrated clustering better separates our two simulated groups.

**3.3 Pathway and Gene Set Overdisperrsion Analysis with Single-Cell RNA-Seq Data**

For a more realistic demonstration, we will analyze single-cell RNA-seq data from Pollen et al. [12]. The error models PAGODA uses are based off of count-based processes and therefore the inputted data will be a matrix of read counts.

1. We can load the read count table and cell group annotations using data("pollen") call. The columns are cells and the rows are genes. Some additional filters are also applied to remove poor cells and non-detected genes. Your own single-cell RNA--seq data can be substituted at this step as well.

```
library(scde)
data(pollen)
# remove poor cells and genes
cd <- clean.counts(pollen)
# check the final dimensions of the read count matrix
dim(cd)
## [1] 11310    64
```

For visualizations later, we will translate group and sample source data from the original publication [12] into color codes.

```
x <- gsub("^Hi_(.*)_.*", "\\1", colnames(cd))
l2cols <- c("coral4", "olivedrab3", "skyblue2",
"slateblue3")[as.integer(factor(x, levels = c("NPC", "GW16", "GW21",
"GW21+3")))]
```

2. Next, we'll construct error models for individual cells. Here, we use a k-nearest neighbor model fitting procedure implemented by knn.error.models() method. This is a relatively noisy dataset (non-UMI), so we raise the min.count.threshold to 2 (minimum number of reads for the gene to be initially classified as a non-failed measurement), requiring at least 5 non-failed measurements per gene. We're providing a rough guess to the complexity of the population, by fitting the error models based on 1/4 of most similar cells (i.e., guessing there might be ~4 subpopulations). Note, this step takes a considerable amount of time unless multiple cores are used. We highly recommend use of multiple cores. You can check the number of available cores available using detectCores().

```
knn <- knn.error.models(cd, k = ncol(cd)/4, n.cores = 1,
min.count.threshold = 2, min.nonfailed = 5, max.model.plots = 10)
```

3. In order to accurately quantify excess variance or overdispersion, we must normalize out expected levels of technical and intrinsic biological noise. Briefly, variance of the NB/Poisson mixture processes derived from the error modeling step is modeled as a chi-squared distribution using adjusted degrees of freedom and observation weights based on the drop-out probability of a given gene. We will normalize variance, trimming 3 most extreme cells and limiting maximum adjusted variance to 5.

```
varinfo <- pagoda.varnorm(knn, counts = cd, trim = 3/ncol(cd),
max.adj.var = 5, n.cores = 1, plot = TRUE)
```

4. Even with all the corrections, sequencing depth or gene coverage is typically still a major aspect of variability. In most studies, we would want to control for that as a technical artifact (exceptions are cell mixtures where subtypes significantly differ in the amount of total mRNA). We will control for the gene coverage (estimated as a number of genes with nonzero magnitude per cell) by normalizing out that aspect of cell heterogeneity:

```
varinfo <- pagoda.subtract.aspect(varinfo, colSums(cd[,
rownames(knn)]>0))
```

5. As mentioned previously, in order to detect significant aspects of heterogeneity across the population of single cells, PAGODA identifies pathways and gene sets that exhibit statistically significant excess of coordinated variability. Specifically, for each gene set, we will test whether the amount of variance explained by the first principal component significantly exceeds the background expectation. We can test both predefined gene sets as well as "de novo" gene sets whose expression profiles are well correlated within the given dataset.

For predefined gene sets, we'll use the GO annotations we previously loaded from liger.

```
# in case you didn't load it previously, load it now
library(liger)
data("org.Hs.GO2Symbol.list")
go.env <- org.Hs.GO2Symbol.list
# remove GOs with too few or too many genes
go.env <- clean.gos(go.env)
# convert to an environment
go.env <- list2env(go.env)
```

Now, we can calculate weighted first principal component magnitudes for each GO gene set in the provided environment and evaluate the statistical significance of their overdispersion.

```
pwpca <- pagoda.pathway.wPCA(varinfo, go.env, n.components = 1, n.cores =
1)
df <- pagoda.top.aspects(pwpca, return.table = TRUE, plot = FALSE,
z.score = 1.96)
head(df)
##            name npc  n    score        z     adj.z
## 339  GO:0003179  1 10 3.495767 11.108780 10.760218
## 338  GO:0003170  1 10 3.495767 11.108780 10.760218
## 3570 GO:0060563  1 12 3.220725 10.643172 10.297292
## 1829 GO:0030426  1 39 3.134488 14.644926 14.338584
## 1302 GO:0014009  1 10 3.105600  9.656705  9.307366
## 1830 GO:0030427  1 40 3.093050 14.530866 14.223476
```

- The z column gives the Z-score of pathway over-dispersion relative to the genome-wide model (Z-score of 1.96 corresponds to *P*-value of 5%, etc.).

  - "z.adj" column shows the Z-score adjusted for multiple hypothesis (using Benjamini-Hochberg correction).

  - "score" gives observed/expected variance ratio.

  - "sh.z" and "adj.sh.z" columns give the raw and adjusted Z-scores of "pathway cohesion," which compares the observed PC1 magnitude to the magnitudes obtained when the observations for each gene are randomized with respect to cells. When such Z-score is high (e.g., for GO:0008009) then multiple genes within the pathway contribute to the coordinated pattern.

6. We can also test "de novo" gene sets whose expression profiles are well correlated within the given dataset. The following procedure will determine "de novo" gene clusters in the data, and build a background model for the expectation of the gene cluster weighted principal component magnitudes. Note the higher trim values for the clusters, as we want to avoid clusters that are formed by outlier cells.

```
clpca <- pagoda.gene.clusters(varinfo, trim = 7.1/ncol(varinfo$mat),
n.clusters = 50, n.cores = 1, plot = FALSE)
```

7. Now the set of top aspects can be recalculated taking these de novo gene clusters into account.

```
df <- pagoda.top.aspects(pwpca, clpca, return.table = TRUE, plot = FALSE,
z.score = 1.96)
head(df)
##                name npc   n    score          z      adj.z
## 339      GO:0003179   1  10 3.495767 11.108780 10.760218
## 338      GO:0003170   1  10 3.495767 11.108780 10.760218
## 4334 geneCluster.8   1 307 3.397680 13.114746 12.814767
## 3570    GO:0060563   1  12 3.220725 10.643172 10.297292
## 1829    GO:0030426   1  39 3.134488 14.644926 14.338584
## 1302    GO:0014009   1  10 3.105600  9.656705  9.307366
```

8. To view top aspects of transcriptional heterogeneity, we will first obtain information on all the significant aspects. We will also determine the overall cell clustering based on this full pathway-level information.

```
# get full info on the top aspects
tam <- pagoda.top.aspects(pwpca, clpca, n.cells = NULL, z.score =
qnorm(0.01/2, lower.tail = FALSE))
# determine overall cell clustering
hc <- pagoda.cluster.cells(tam, varinfo)
```

9. We can then reduce redundant aspects in two steps. In the first step, we will combine pathways that are driven by the same sets of genes. In the second step we will combine aspects that show similar patterns (i.e., separate the same sets of cells).

```
tamr <- pagoda.reduce.loading.redundancy(tam, pwpca, clpca)
tamr2 <- pagoda.reduce.redundancy(tamr, distance.threshold = 0.9, trim =
0, plot = FALSE)
```

10. We can then view these top aspects in a heatmap (Fig. 7). Indeed, we see a correspondence between out derived cell annotations and the previously published annotations.

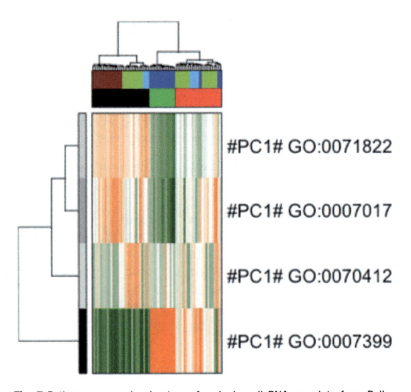

**Fig. 7** Pathway expression heatmap for single-cell RNA-seq data from Pollen et al. The columns are cells and the rows represent a cluster of pathways. The row names are assigned to be the top overdispersed aspect in each cluster. The green-to-orange color scheme shows low-to-high weighted PC scores (aspect patterns), where generally orange indicates higher expression and green lower expression. The column colors are cell annotations from the original publication

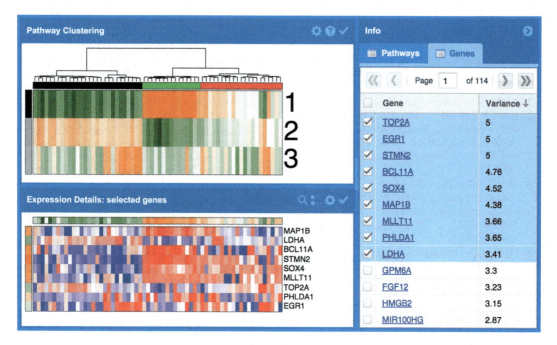

**Fig. 8** Sample screenshot of an interactive PAGODA app

```
col.cols <- rbind(groups = cutree(hc, 3), l2cols)
pagoda.view.aspects(tamr2, cell.clustering = hc, box = TRUE, labCol = NA,
margins = c(0.5, 20), col.cols = rbind(col.cols))
```

> 11. To interactively browse and explore the output, we can also create a PAGODA app (Fig. 8).

```
# compile a browsable app, showing top three clusters with the top color
bar
app <- make.pagoda.app(tamr2, tam, varinfo, go.env, pwpca, clpca,
col.cols = col.cols, cell.clustering = hc, title = "NPCs")
# show app in the browser (port 1468)
show.app(app, "pollen", browse = TRUE, port = 1468)
```

> The PAGODA app allows you to view the gene sets grouped within each aspect (row), as well as genes underlying the detected heterogeneity patterns. In this manner, you can interactively explore the pathways and genes driving each identified transcriptional subpopulation.

## Acknowledgment

This work was supported by NIH grant F99CA222750.

## References

1. Soneson C, Delorenzi M (2013) A comparison of methods for differential expression analysis of RNA-seq data. BMC Bioinformatics 14:91. https://doi.org/10.1186/1471-2105-14-91

2. Jaakkola MK, Seyednasrollah F, Mehmood A, Elo LL (2016) Comparison of methods to detect differentially expressed genes between single-cell populations. Brief Bioinform 18: bbw057. https://doi.org/10.1093/bib/bbw057

3. Kharchenko PV, Silberstein L, Scadden DT (2014) Bayesian approach to single-cell differential expression analysis. Nat Methods 11:740–742. https://doi.org/10.1038/nmeth.2967

4. Finak G, McDavid A, Yajima M, Deng J, Gersuk V, Shalek AK, Slichter CK, Miller HW, McElrath MJ, Prlic M, Linsley PS, Gottardo R (2015) MAST: a flexible statistical framework for assessing transcriptional changes and characterizing heterogeneity in sin gle-cell RNA sequencing data. Genome Biol 16:278. https://doi.org/10.1186/s13059-015-0844-5

5. Subramanian A, Tamayo P, Mootha VK, Mukherjee S, Ebert BL, Gillette MA, Paulovich A, Pomeroy SL, Golub TR, Lander ES, Mesirov JP (2005) Gene set enrichment analysis: a knowledge-based approach for interpreting genome-wide expression profiles. Proc Natl Acad Sci U S A 102:15545–15550. https://doi.org/10.1073/pnas.0506580102

6. Fan J, Salathia N, Liu R, Kaeser GE, Yung YC, Herman JL, Kaper F, Fan J-B, Zhang K, Chun J, Kharchenko PV (2016) Characterizing transcriptional heterogeneity through pathway and gene set overdispersion analysis. Nat Methods 13:241–244. https://doi.org/10.1038/nmeth.3734

7. Wagner F (2016) The XL-mHG test for enrichment: algorithms, bounds, and power. https://doi.org/10.7287/peerj.preprints.1962v1

8. Huang DW, Sherman BT, Lempicki RA (2009) Bioinformatics enrichment tools: paths toward the comprehensive functional analysis of large gene lists. Nucleic Acids Res 37:1–13. https://doi.org/10.1093/nar/gkn923

9. R Core Team (2017) R: a language and environment for statistical computing

10. Liberzon A, Birger C, Thorvaldsdóttir H, Ghandi M, Mesirov JP, Tamayo P (2015) The molecular signatures database (MSigDB) hallmark gene set collection. Cell Syst 1:417–425. https://doi.org/10.1016/j.cels.2015.12.004

11. Dunnett CW (1955) A multiple comparison procedure for comparing several treatments with a control. J Am Stat Assoc 50:1096–1121. https://doi.org/10.1080/01621459.1955.10501294

12. Pollen AA, Nowakowski TJ, Shuga J, Wang X, Leyrat AA, Lui JH, Li N, Szpankowski L, Fowler B, Chen P, Ramalingam N, Sun G, Thu M, Norris M, Lebofsky R, Toppani D, Kemp DW, Wong M, Clerkson B, Jones BN, Wu S, Knutsson L, Alvarado B, Wang J, Weaver LS, May AP, Jones RC, Unger MA, Kriegstein AR, West JAA (2014) Low-coverage single-cell mRNA sequencing reveals cellular heterogeneity and activated signaling pathways in developing cerebral cortex. Nat Biotechnol 32:1053–1058. https://doi.org/10.1038/nbt.2967

# Chapter 8

## Pseudotime Reconstruction Using TSCAN

### Zhicheng Ji and Hongkai Ji

## Abstract

In many single-cell RNA-seq (scRNA-seq) experiments, cells represent progressively changing states along a continuous biological process. A useful approach to analyzing data from such experiments is to computationally order cells based on their gradual transition of gene expression. The ordered cells can be viewed as samples drawn from a pseudo-temporal trajectory. Analyzing gene expression dynamics along the pseudotime provides a valuable tool for reconstructing the underlying biological process and generating biological insights. TSCAN is an R package to support in silico reconstruction of cells' pseudotime. This chapter introduces how to apply TSCAN to scRNA-seq data to perform pseudotime analysis.

**Key words** Single-cell RNA-seq, Gene expression, Pseudotime, Minimum spanning tree, Genomics, Bioinformatics

## 1  Introduction

Single-cell RNA-seq (scRNA-seq) offers unprecedented power for analyzing cells' distinct transcriptomic profiles in a heterogeneous cell population [1–3]. In many studies, a biological sample consists of cells from different stages of a biological process. For example, during cell differentiation, cells may differentiate at different speeds and they can enter different developmental lineages. As a result, when a sample is obtained at a particular time point of a differentiation process, the sample may contain cells representing different developmental stages and lineages. A useful approach to utilizing scRNA-seq data obtained from such a biological sample is to computationally place cells onto a pseudo-temporal trajectory based on their progressive changes of gene expression. Such a pseudo-temporal trajectory often reflects the underlying biological process from which the cells are sampled. Ordering cells along this pseudo-temporal trajectory and analyzing cells' transcriptomic changes along the pseudotime therefore could yield new insights into the dynamic gene expression and regulation programs of a biological process. This approach, also known as "pseudotime analysis," has

Guo-Cheng Yuan (ed.), *Computational Methods for Single-Cell Data Analysis*, Methods in Molecular Biology, vol. 1935, https://doi.org/10.1007/978-1-4939-9057-3_8, © Springer Science+Business Media, LLC, part of Springer Nature 2019

been emerging as a powerful tool for studying cell differentiation, immune response, tumor progression, and many other biological processes [1, 4–7].

A pioneering work that demonstrates the value of pseudotime analysis in scRNA-seq experiments is [1]. In that study, a computational algorithm Monocle was proposed to construct pseudo-time using minimum spanning tree (MST). Since then, a number of pseudotime reconstruction methods have been developed. Comprehensive reviews of such methods can be found in [8, 9]. This chapter introduces one such method, TSCAN [6], which is developed by us to support pseudotime analysis. In several recent benchmark studies, TSCAN has shown favorable performance compared to other available methods [8, 10].

Similar to Monocle [1], TSCAN uses MST to construct pseudo-temporal trajectory. However, instead of treating each cell as a tree node, TSCAN first groups similar cells into clusters. It then treats each cluster as a node and constructs an MST to connect cluster centers. The resulting MST provides the backbone of the pseudo-temporal path. Individual cells are then projected to the tree backbone to determine their pseudotime and order on the path (Fig. 1).

Unlike TSCAN, Monocle directly uses individual cells rather than cell clusters as tree nodes to construct MST. A drawback of this approach is the instability of the tree inference. When there are a large number of cells, the tree space is highly complex due to the large number of tree nodes. Random noises can easily change the topology of the MST, making the tree inference highly variable and unstable [6]. As a result, the MST obtained by this approach often deviates from the true biology. Moreover, the computation also becomes more challenging with a large number of cells. By clustering cells and treating clusters as nodes, TSCAN substantially reduces the number of tree nodes and hence the complexity of the tree space, making the tree inference less variable and more stable. Similar to the variance-bias tradeoff in machine learning, the reduced complexity of the tree space often results in improved tree estimation. A systematic evaluation in [6] shows that by using this clustered-based MST approach, TSCAN was better able to reconstruct the underlying biological processes compared to Monocle.

TSCAN is freely available as an open-source R package. In the following sections, we will introduce how TSCAN can be used to perform pseudotime analysis step by step.

## 2    Materials

In order to use TSCAN, one needs to have the following software and data.

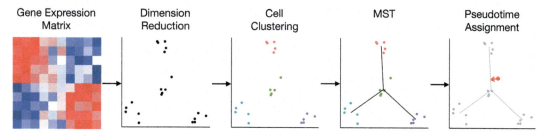

**Fig. 1** TSCAN analysis workflow. Starting from a preprocessed gene expression matrix, TSCAN constructs cells' pseudo-temporal ordering using the following steps: reduce dimension of the gene expression matrix, group cells into clusters, construct MST that links the cluster centers to create the backbone of the pseudo-temporal trajectory, and project individual cells onto tree edges to obtain pseudotime

**2.1 R**

R is an open-source software for statistical computing that can be freely downloaded and installed [11].

**2.2 TSCAN Package**

TSCAN is an open-source R package that can be freely downloaded and installed from Bioconductor [12] or Github [13]. After installing R, the latest TSCAN can be installed from Github by typing the following commands in R:

```
if (!require("devtools"))
    install.packages("devtools")
devtools::install_github("zji90/TSCAN")
```

**2.3 ScRNA-Seq Data**

We assume that raw scRNA-seq data have already been processed and summarized into a matrix of normalized gene expression values (*see* **Note 1**). In this matrix, each row represents a gene and each column represents a cell. The matrix is stored in a tab-delimited text file. The first row of the file stores cell names, and the first column contains gene names.

In this chapter, we will demonstrate the pseudotime analysis using a scRNA-seq dataset of differentiating human skeletal muscle myoblasts [1]. The dataset contains 271 cells collected at 0, 24, 48, and 72 h after switching human myoblasts to low serum. The log2 transformed gene expression matrix (HSMM_Y) can be downloaded from Github [14].

# 3 Methods

A typical pseudotime analysis using TSCAN consists of four steps: (1) loading data and preprocessing, (2) dimension reduction and cell clustering, (3) pseudotime reconstruction, (4) differential gene expression analysis.

**3.1 Loading Data and Preprocessing**

The first step of TSCAN is to load the scRNA-seq data into R and convert it into a matrix object. For example, the HSMM dataset, which is stored in a tab-delimited text file, can be loaded using the following R command (*see* **Note 2**):

```
library(TSCAN)
data <- as.matrix(read.table("HSMM_Y.txt", as.is=T, header=T, sep="\t",
row.names=1))
```

Next, the data can be preprocessed in multiple ways using the function "preprocess" provided by TSCAN. Using this function, one can log transform the expression matrix and remove genes with low expression or low variability. A gene with low expression can be defined by comparing its mean expression value across cells to a user-specified cutoff, or by comparing the percentage of cells with nonzero expression to a cutoff. Low variability is defined by comparing a gene's coefficient of variation across cells to a user-specified cutoff. ScRNA-seq data can be sparse and contain many zero gene expression values. To mitigate sparsity, one can also use the "preprocess" function to group co-expressed genes into clusters. For each gene cluster, the mean expression of all genes in the cluster will be computed for each cell. By aggregating information from multiple genes, the cluster-level expression typically is more continuous and much less sparse. The following R command demonstrates preprocessing using the HSMM data:

```
aggdata <- preprocess(data, clusternum=0.05*nrow(data), takelog=FALSE,
minexpr_value=0, minexpr_percent=0, cvcutoff=0)
```

This command removes genes that have zero expression in all cells. It also groups genes into clusters and returns the mean gene expression for each cluster. Here the cluster number is equal to 5% of the total gene number. The cluster-level expression is stored in a new matrix `aggdata` which will be used for subsequent pseudo-time reconstruction.

**3.2 Dimension Reduction and Cell Clustering**

In order to construct pseudo-temporal trajectory, TSCAN first groups similar cells into clusters and then constructs a minimum spanning tree (MST) to connect cluster centers. Compared to constructing an MST to connect individual cells, using cell clusters as tree nodes in the MST can reduce the complexity of the tree space and improve the accuracy and stability of tree inference [6].

The default procedure used by TSCAN to cluster cells involves two steps. In the first step, TSCAN reduces the dimension of the input data using principal component analysis (PCA) (*see* **Note 3**). Dimension reduction will help visualization and mitigate the curse of dimensionality in computing cells' distances. The number of principal components (PCs) to keep is automatically determined using an elbow method [6]. In the second step, model-based clustering [15] is used to cluster cells based on the dimension-reduced data (*see* **Note 4**). The number of clusters is automatically determined using the Bayesian information criteria (BIC). These two steps can be carried out using a single R function, as illustrated below using the HSMM data:

```
HSMMmclust <- exprmclust(aggdata)
```

One can visualize the dimension-reduced data and cell cluster-ing results. For example, the following R function generates a scatterplot (Fig. 2a) that visualizes the second and third PCs as well as cell clustering:

```
plotmclust(HSMMmclust, x=2, y=3)
```

In this plot, each dot represents a cell. Different cell clusters are represented using different colors and marker shapes. Cluster cen-ters are marked with numbers.

**3.3 Pseudotime Reconstruction**

Once cells are clustered, TSCAN will construct an MST to connect the cluster centers. This tree will serve as the backbone of the pseudo-temporal trajectory. The tree may have multiple branches. By default, TSCAN will choose the path with the largest number of cell clusters as the main pseudo-temporal path. If two paths have the same number of cell clusters, the path with the largest number of cells will be chosen as the main path. For example, the main path in Fig. 2a will be path 3-1-4.

Sometimes, some branches of the tree are not of biological interest. For example, some cell clusters and tree branches represent contaminated cells. To help identify and remove such branches, one can use TSCAN to visualize expression of marker genes. To dem-onstrate, the following R commands show the expression of two marker genes in the HSMM data: PDGFRA (Fig. 2b) and MYOG (Fig. 2c).

```
plotmclust(HSMMmclust, x=2, y=3,
markerexpr=data["ENSG00000134853.7",],showcluster = F)
plotmclust(HSMMmclust, x=2, y=3,
markerexpr=data["ENSG00000122180.4",],showcluster = F)
```

This dataset contains some contaminated cells which are marked by high expression of PDGFRA [1]. On the other hand, MYOG is a key marker gene for human skeletal muscle cell differ-entiation, and its expression is expected to increase as cells differen-tiate [1]. Based on this prior knowledge, cell cluster 3 is most likely contaminated cells because it has high PDFGRA expression. Clus-ter 2 has relatively low expression of MYOG, and cluster 4 has

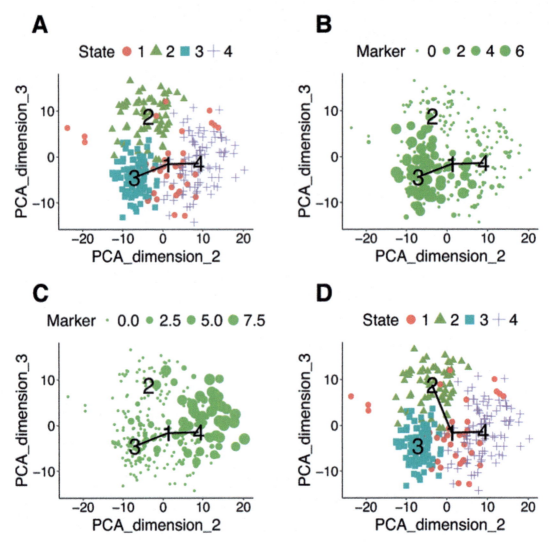

**Fig. 2** TSCAN analysis of the HSMM data. A. Scatterplot showing PC2, PC3, and cell clustering. Path of interest is 3-1-4. B. Visualization of PDGFRA marker gene expression. C. Visualization of MYOG marker gene expression. D. Scatterplot showing PC2, PC3, and cell clustering. Path of interest is 2-1-4

relatively high MYOG expression. Thus, one can identify cluster 2 as undifferentiated cells and cluster 4 as differentiated cells. Since the primary interest here is to study human skeletal muscle cell differentiation, the pseudo-temporal path of main interest should be 2-1-4 rather than the default main path 3-1-4. Using the following R command, TSCAN allows one to choose a path (e.g., path 2-1-4) other than the default main path for visualization and analysis (Fig. 2d):

```
plotmclust(HSMMmclust, x=2, y=3, MSTorder=c(2,1,4))
```

After obtaining the backbone of pseudo-temporal trajectory using the cluster-based MST, one can then order cells along the pseudo-temporal path and assign pseudotime to cells. TSCAN orders cells in two steps. First, cell clusters are ordered based on the pseudo-temporal path. For example, for path 2-1-4, all cells in cluster 2 are placed before cells in cluster 1, and cells in cluster 1 are placed before cells in cluster 4. Second, in order to order cells within the same cluster, cells are projected onto the edge that connects the centers of neighboring clusters, and the projection determines cell ordering (Fig. 1, Pseudotime Assignment). For example, cells in cluster 2 are projected onto the edge linking the centers of cluster 2 and cluster 1. Cells with a projection closer to the center of cluster 2 are placed before cells with a projection closer to the center of cluster 1. Once cell ordering is determined, the order of each cell is used as its pseudotime. As an example, the following command can be used to reconstruct cells' pseudo-temporal ordering along the path 2-1-4:

```
HSMMTSCANorder214 <- TSCANorder(HSMMmclust, MSTorder=c(2,1,4))
```

After pseudotime reconstruction, one can visualize how gene expression changes along pseudotime. For example, the following command will generate a plot to show how the expression of MYH2 changes along the pseudotemporal path 2-1-4 (Fig. 3a):

```
singlegeneplot(data["ENSG00000125414.13",],HSMMTSCANorder214)
```

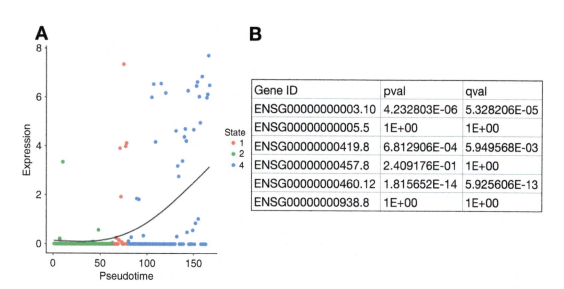

**Fig. 3** Differential gene detection by TSCAN. (**a**) Scatterplot showing expression of MYH2 gene along the pseudotemporal path 2-1-4. (**b**) An example of TSCAN output that shows a few top differentially expressed genes

Since the pseudo-temporal trajectory constructed using MST may contain multiple branches, one can use the following command to obtain the pseudo-temporal ordering of all possible paths (*see* **Note 5**):

```
HSMMTSCANorderfull <- TSCANorder(HSMMmclust,listbranch = T)
```

***3.4 Differential Gene Expression Analysis***

Given cells' pseudo-temporal ordering, one can detect genes with significant expression changes along pseudotime. To detect such genes, TSCAN fits a generalized additive model (GAM) for each gene to describe the functional relationship between gene expression and pseudotime. The fitted model is compared to a null model that assumes constant expression along pseudotime. The model fitting is performed using mgcv package in R [16]. A likelihood ratio test is conducted to obtain $p$-value. To account for multiple testing, $p$-values are converted to false discovery rate (FDR). By default, genes with FDR $< 0.05$ are reported as differential. To demonstrate, the following command performs differential analysis along the pseudo-temporal path 2-1-4:

```
diffres <- difftest(data,HSMMTSCANorder214)
head(differs)
```

The returned object is a data frame that contains the $p$-values and FDRs for differential tests (Fig. 3b).

# 4   Notes

1. There are a number of different technology platforms to generate scRNA-seq data. Data from different platforms can have very different characteristics. Currently, there is no universal computational pipeline that can optimally process all types of scRNA-seq data. Thus, while TSCAN provides pseudotime analysis functions, it is users' own responsibility to find the most appropriate way to map sequence reads and convert the read data into normalized gene expression matrices before pseudotime analysis.

2. The matrix of normalized gene expression values used as TSCAN input can also be stored in text files that use other common delimiters. For example, instead of using a tab-delimited text file, one can store the input data in a comma-separated value (CSV) file that uses comma to separate different columns. To load such data, one only needs to change the separator (i.e., the "sep=" argument) in the read.table command. For example:

```
data <- as.matrix(read.table("HSMM_Y.csv", as.is=T, header=T, sep=",",
row.names=1))
```

3. While TSCAN uses PCA as its default dimension reduction method, there are many other methods in R that can be used for dimension reduction (e.g., tsne [17]). TSCAN provides users with the flexibility to use a dimension reduction method of their choice. In order to do so, users only need to run dimension reduction using the R function they choose. Suppose the dimension-reduced data are stored in matrix `customdimred`. Users can supply this matrix to TSCAN for clustering cells by the following command:

```
HSMMmclust <- exprmclust(customdimred, reduce=F)
```

4. TSCAN also provides users with the flexibility to choose how to cluster cells. Instead of using model-based clustering, users can supply their own clustering results derived by other clustering methods such as k-means clustering or hierarchical clustering. The following command accepts both customized dimension reduction and cell clustering results:

```
HSMMmclust <- exprmclust(customdimred, reduce=F, cluster=customcluster)
```

After loading the cell clustering results, users can then proceed with the remaining steps of pseudotime reconstruction.

5. For user's convenience, TSCAN also provides a graphical user interface (GUI) to perform pseudotime analysis. Most of the functions and commands described above have corresponding buttons in the GUI. Instead of typing the commands in R, one can also use GUI to execute the same functions. This usually only requires one to click a few buttons. The link to the online version of TSCAN GUI can be found on TSCAN's Github page [13]. On that page, there is a video demonstrating how to use the GUI. Since the video is quite straightforward, we will not repeat the demonstration of GUI here. The TSCAN GUI can also be invoked locally in R on user's own computer using the following command:

```
TSCANui()
```

Compared to the GUI, using the R commands described in this chapter can give users more control of the analysis. Moreover, these R commands can serve as building blocks of users' own analysis pipelines and can be easily incorporated into their codes and programs.

## References

1. Trapnell C, Cacchiarelli D, Grimsby J et al (2014) The dynamics and regulators of cell fate decisions are revealed by pseudotemporal ordering of single cells. Nat Biotechnol 32 (4):381–386

2. Zheng C, Zheng L, Yoo JK et al (2017) Landscape of infiltrating T cells in liver cancer revealed by single-cell sequencing. Cell 169 (7):1342–1356

3. Shalek AK, Satija R, Shuga J et al (2014) Single-cell RNA-seq reveals dynamic paracrine control of cellular variation. Nature 510 (7505):363–369

4. Marco E, Karp RL, Guo G et al (2014) Bifurcation analysis of single-cell gene expression data reveals epigenetic landscape. Proc Natl Acad Sci U S A 111(52):E5643–E5650

5. Shin J, Berg DA, Zhu Y et al (2015) Single-cell RNA-Seq with waterfall reveals molecular cascades underlying adult neurogenesis. Cell Stem Cell 17(3):360–372

6. Ji Z, Ji H (2016) TSCAN: pseudo-time reconstruction and evaluation in single-cell RNA-seq analysis. Nucleic Acids Res 44(13):e117–e117

7. Haghverdi L, Buettner M, Wolf FA et al (2016) Diffusion pseudotime robustly reconstructs lineage branching. Nat Methods 13 (10):845–848

8. Wouter S, Robrecht C, Helena T, et al (2018) A comparison of single-cell trajectory inference methods: towards more accurate and robust tools. bioRxiv. https://doi.org/10.1101/276907

9. Herring CA, Chen B, McKinley ET et al (2018) Single-cell computational strategies for lineage reconstruction in tissue systems. Cell Mol Gastroenterol Hepatol 5(4):539–548

10. Street K, Risso D, Fletcher RB et al (2018) Slingshot: cell lineage and pseudotime inference for single-cell transcriptomics. BMC Genomics 19(1):477

11. R project. https://www.r-project.org/

12. TSCAN R package on Bioconductor. https://www.bioconductor.org/packages/release/bioc/html/TSCAN.html

13. TSCAN R package on Github. https://github.com/zji90/TSCAN

14. HSMM singe-cell RNA-seq dataset. https://raw.githubusercontent.com/zji90/TSCANdata/master/HSMM_Y.txt

15. Fraley C, Raftery AE (2002) Model-based clustering, discriminant analysis, and density estimation. J Am Stat Assoc 97(458):611–631

16. Wood SN (2011) Fast stable restricted maximum likelihood and marginal likelihood estimation of semiparametric generalized linear models. J Royal Stat Soc Sec B 73(1):3–36

17. Maaten LVD, Hinton G (2008) Visualizing data using t-SNE. J Mach Learn Res 9:2579–2605

# Chapter 9

## Estimating Differentiation Potency of Single Cells Using Single-Cell Entropy (SCENT)

### Weiyan Chen and Andrew E. Teschendorff

### Abstract

The ability to measure molecular properties (e.g., mRNA expression) at the single-cell level is revolutionizing our understanding of cellular developmental processes and how these are altered in diseases like cancer. The need for computational methods aimed at extracting biological knowledge from such single-cell data has never been greater. Here, we present a detailed protocol for estimating differentiation potency of single cells, based on our Single-Cell ENTropy (SCENT) algorithm. The estimation of differentiation potency is based on an explicit biophysical model that integrates the RNA-Seq profile of a single cell with an interaction network to approximate potency as the entropy of a diffusion process on the network. We here focus on the implementation, providing a step-by-step introduction to the method and illustrating it on a real scRNA-Seq dataset profiling human embryonic stem cells and multipotent progenitors representing the 3 main germ layers. SCENT is aimed particularly at single-cell studies trying to identify novel stem-or-progenitor like phenotypes, and may be particularly valuable for the unbiased identification of cancer stem cells. SCENT is implemented in R, licensed under the GNU General Public Licence v3, and freely available from https://github.com/aet21/SCENT.

**Key words** Single-cell, RNA-Seq, Differentiation potency, Network, Entropy

## 1 Introduction

Over 50 years ago Conrad Waddington proposed an epigenetic landscape model of cellular differentiation, in which cell-fate transitions are modeled as canalization events, with stable cell states occupying the basins or attractor states [1, 2]. A key ingredient of this landscape is the energy potential or height [3], which correlates with cell-potency, i.e., the number of cell-fate choices a given cell may have. According to this landscape model of cellular differentiation, human embryonic stem cells (hESCs), owing to their pluripotency, occupy the highest attractor state, with terminally differentiated cells occupying the lowest lying basins. In between these extremes and within a specific cell-lineage may lie other attractor states representing progenitor cells of variable degrees of

Guo-Cheng Yuan (ed.), *Computational Methods for Single-Cell Data Analysis*, Methods in Molecular Biology, vol. 1935, https://doi.org/10.1007/978-1-4939-9057-3_9, © Springer Science+Business Media, LLC, part of Springer Nature 2019

potency. Although Waddington's landscape model is an unrealistic oversimplification, with newer proposed models [4] indicating substantially more intercellular heterogeneity and higher-order bifurcations, it is clear that quantifying the potency of single cells remains a task of critical importance: quantifying single-cell potency can not only help us understand cell-type heterogeneity within FACS-sorted cell populations and allow explicit construction of cell differentiation landscapes [2, 5], but importantly, can also provide us with an unbiased means of identifying novel stem and progenitor cell phenotypes. For instance, estimating potency of cells may be particularly important in the context of cancer, where one may wish to identify putative cancer stem cell phenotypes, which are thought to drive tumor growth, tumor heterogeneity, and drug resistance [6, 7].

It is reasonable to assume that potency is encoded by the genome-wide transcriptomic profile of the cell, and recent technological advances in single-cell RNA-Seq data generation [3, 8, 9] are allowing us to test, for the first time, explicit biophysical models for estimating differentiation potency of individuals cells. We note that the task of estimating differentiation potency of single cells is different to that of most single-cell analysis algorithms proposed so far [10–18], whose main aim is to infer cell-lineage differentiation trajectories, typically in the context of time-course scRNA-Seq data. Although these algorithms infer pseudotime, which in time-course differentiation experiments can be interpreted as differentiation potency, these tools do not use an explicit biophysical model to estimate it, as they often involve a process of training or feature selection that draws information from all cells in the study. Alternatively, the tools might require prior biological knowledge (e.g., surface marker expression) as to which cells may define start or root nodes. As such, most of these tools do not represent general models and therefore are not applicable for estimating potency, for instance in the context of non-time-course scRNA-Seq data [19] or scRNA-Seq data from cancer tissue [6, 7], or may only be applicable within a specific lineage (e.g., the hematopoietic lineage). Indeed, so far only a few explicit biophysical models for estimating differentiation potency of single cells have been proposed. These are (1) StemID [20], an algorithm that approximates differentiation potency by computing a genome-wide transcriptomic entropy measuring how uniformly expressed the genes are, (2) SLICE [21], a method that defines potency in terms of the Shannon entropy over gene expression derived gene-ontology activity estimates, in effect measuring the uniformity of the activation profile of different biological processes in a cell, and (3) Signaling Entropy or Single-Cell Entropy (SCENT) [22], which models potency in terms of the entropy-rate of a diffusion process [23] on a signaling network, thus measuring how efficiently signaling can diffuse over the whole network. A recent comparative study of these methods has shown that the Signaling Entropy measure used in SCENT currently offers the most reliable means of estimating differentiation potency [22], and importantly its validity has been widely demonstrated not only on single cells but also on bulk tissue samples and cell-lines [24, 25]. Moreover, the Signaling Entropy measure used to approximate differentiation potency has been shown to correlate with drug resistance [25] and clinical outcome in lung and breast cancer [26]. For these reasons, this chapter focuses on the estimation of differentiation potency via the Signaling Entropy method used in the SCENT package. As we shall see below, we assume that an RNA-Seq dataset (be it single cell or bulk sample) has been normalized at the intra cell/sample and inter cell/sample levels, meaning that batch effects or other technical confounders have been accounted for. However, as we stress again later, the signaling entropy calculation is fairly robust to the normalization procedure and fairly insensitive to batch effects, as it only depends on ratios of gene expression, not their absolute levels.

## 2 Materials

### 2.1 Software and Hardware Required

The estimation of differentiation potency using SCENT requires the following hardware and software:

1. A standard desktop or laptop computer with Windows, Max OSX, or Linux operating system. We recommend at least 16GB RAM.
2. R statistical computing environment (version 3.4.3 or later) from The Comprehensive R Archive Network (https://cran.r-project.org/).
3. Install R and Bioconductor packages into the R environment, including Biobase, mclust, igraph, isva, cluster, corpcor. We also recommend installing the package parallel to speed up computations.
4. Download and install SCENT R-package from https://github.com/aet21/SCENT.

### 2.2 RNA-Seq Data

The first data input for the computation of differentiation potency using the SCENT package is a single-cell RNA-Seq dataset. The procedure is identical for bulk RNA-Seq data, the only difference being in the specific preprocessing and normalization of the data. To illustrate the method, we shall use a scRNA-Seq dataset from Chu et al. [19], generated with the Fluidigm C1 platform. This dataset can be downloaded from the GEO website under accession number GSE75748, i.e., via *https://www.ncbi.nlm.nih.gov/geo/query/acc.cgi?acc=GSE75748*, and the specific file to download is *GSE75748 sc cell type ec.csv.gz*. It contains scRNA-Seq profiles for

**Table 1**
**Distribution and number of single-cell types in Chu et al. dataset**

| Index label | | Cell-type | Potency | Number (after QC) |
|---|---|---|---|---|
| 1 | hESC | Human embryonic stem cell | Pluripotent | 374 |
| 2 | NPC | Ectoderm progenitor | Non-pluripotent | 173 |
| 6 | DEC | Endoderm progenitor | Non-pluripotent | 138 |
| 5 | EC | Mesoderm progenitor | Non-pluripotent | 105 |
| 4 | HFF | Human foreskin fibroblasts | Non-pluripotent | 159 |
| 3 | TB | Trophoblasts | Non-pluripotent | 69 |

1018 single cells, composed of 374 human embryonic stem cells, 173 neural progenitor cells (NPCs), 138 definite endoderm progenitors (DEPs), 105 mesoderm progenitors of endothelial cells (ECs), 69 trophoblasts (TBs), and 159 human fibroblasts (HFFs), as indicated in Table 1. These cells were obtained via FACS-sorting and/or induced differentiation experiments from hESCs [19]. We also provide a log-normalized version of the scRNA-Seq data from the github link *https://github.com/aet21/SCENT/* under filename "sceChu.Rd", for easy uploading into R.

*2.3 User-Defined Functional Gene Network*

The second required input for the SCENT algorithm is a user-defined functional gene network, for instance, a protein-protein interaction (PPI) network documenting the main interactions that take place in a cell. For justification as to why a PPI network is needed (*see* **Notes 1** and **2**). Although these networks are mere caricatures of the underlying signaling networks, ignoring time, spatial, and biological contexts, some features of the network may nevertheless be informative, and as we shall see below this is indeed the case. For instance, if protein-A is a hub (a node of very high connectivity) and protein-B has only a few connections, then it is likely that protein-A will have a higher connectivity than protein-B in any particular biological context (unless of course protein-A is absent from the cell altogether). As we shall later, the scRNA-Seq data will provide us with the biological context in which to generate context or cell-type-specific networks. The specific PPI network we use here is derived from Pathway Commons (www.pathwaycommons.org) (downloaded in Jan. 2016), which is an integrated resource collating together PPIs from several distinct sources. In particular, the network is constructed by integrating the following sources: the Human Protein Reference Database (HPRD), the National Cancer Institute Nature Pathway Interaction Database (NCI-PID) (http://pid.nci.nih.gov), the Interactome (Intact), and the Molecular Interaction Database (MINT). Protein interactions in this network include physical stable interactions such as those defining protein complexes, as well as transient interactions such as posttranslational modifications and enzymatic reactions found in signal transduction pathways. We focused on nonredundant interactions, only included nodes with an Entrez gene ID annotation, and focused on the maximally connected component (*see* **Note 3**), resulting in a connected network of 10,720 nodes (unique Entrez IDs) and 152,889 documented interactions. This network can be downloaded from *https://github.com/aet21/SCENT/* under filename *ppiAsigH-PC2-17Jan2016.Rd*. Another earlier version of the network of size 8434 nodes can be found under filename *hprdAsigH-13Jun12.Rd*.

Here, we encode the network as an adjacency matrix of dimension $8434 \times 8434$, with "0" entries indicating that no interaction or connection between the two genes has been documented, while a "1" means that an interaction has been documented. Diagonal entries are set to 0. Importantly, because SCENT will integrate the network with RNA-Seq data, the row names and column names of the network must use the same gene identifier as used in the RNA-Seq data matrix.

# 3 Methods

## 3.1 Workflow for Signaling Entropy Calculation

The estimation of differentiation potency with SCENT consists of three major steps: (1) checking and further processing (if required) of the normalized scRNA-Seq data, (2) integration of the scRNA-Seq data with the user-defined gene functional network, (3) computation of the Signaling Entropy Rate (denoted *SR*) which is used to approximate differentiation potency of single cells (Fig. 1a). Optionally, SCENT can also be used to quantify the heterogeneity in potency of a single-cell population (Fig. 1b, c), and to infer lineage relationships between the major clusters of single-cells [22]. However, because SCENT was designed mainly for the estimation of differentiation potency we shall only focus on the method for estimating it and how it can be used to infer discrete potency states.

### 3.1.1 Checking and Further Preprocessing of the scRNA-Seq Data

We assume that the scRNA-Seq data have been appropriately normalized. Given a count matrix of reads mapping to genes, we assume that the user has run this count matrix through a single-cell processing and quality control pipeline (*see*, e.g., [27]), such as that provided by the Bioconductor package *scater* [28] or R-package *Seurat* [29]. The end result of this is typically a log-normalized scRNA-Seq data matrix. The log-transformation provides a natural regularization of the data, stabilizing the variance of highly expressed genes, and is strongly recommended for any down-stream analyses [27], especially in the context of SCENT. In addition, we must also take care of the smallest values in the

# Flowchart for S̲ingle-C̲ell E̲ntropy (SCENT)

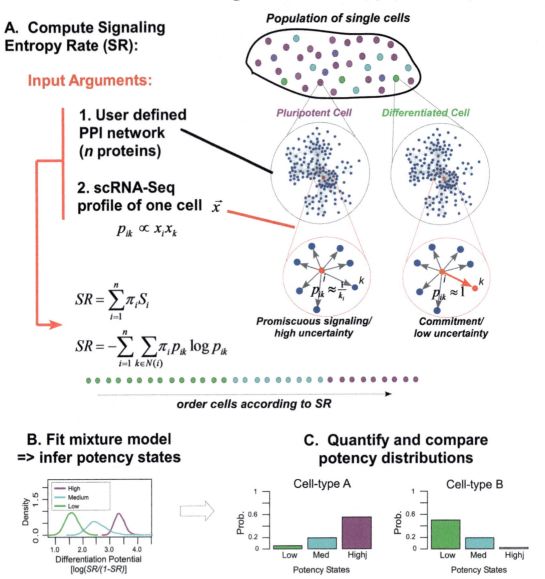

**A. Compute Signaling Entropy Rate (SR):**

**Input Arguments:**

**1. User defined PPI network (*n* proteins)**

**2. scRNA-Seq profile of one cell** $\vec{x}$

$$p_{ik} \propto x_i x_k$$

$$SR = \sum_{i=1}^{n} \pi_i S_i$$

$$SR = -\sum_{i=1}^{n} \sum_{k \in N(i)} \pi_i p_{ik} \log p_{ik}$$

*Population of single cells*

*Pluripotent Cell*    *Differentiated Cell*

$p_{ik} \approx \frac{F}{k_i}$    $k$

$p_{ik} \approx 1$    $k$

*Promiscuous signaling/ high uncertainty*    *Commitment/ low uncertainty*

*order cells according to SR*

**B. Fit mixture model => infer potency states**

Density

High
Medium
Low

1.0  2.0  3.0  4.0
Differentiation Potential
[log(SR/(1−SR))]

**C. Quantify and compare potency distributions**

Cell-type A

Prob.
1
0.6
0.2
0

Low  Med  Highj
Potency States

Cell-type B

Prob.
1
0.6
0.2
0

Low  Med  Highj
Potency States

**Fig. 1** (a) The estimation of differentiation potency using the signaling entropy rate requires two inputs: a user-defined gene functional network (e.g., a PPI network) which does not depend on biological context, and a normalized scRNA-Seq profile of a cell or sample, which thus provides the biological context. The latter profile is overplayed onto the network to define a cell-specific stochastic matrix $P$ with entries $p_{ij}$. From this matrix, we can derive the invariant measure (steady-state probability) $\pi_i$, which satisfies $\pi P = \pi$, and finally the signaling entropy rate $SR$ is obtained as a weighted average over local signaling entropies. This allows cells to be ordered according to their $SR$ values, i.e., according to potency. $SR$ can be quantified on a scale between 0 and 1 (not shown). (b) Transforming the normalized $SR$ to the logit-scale and fitting Gaussian Mixture models allows the identification of potency states. (c) The distribution of these potency states across cell-types can be analyzed to identify novel cellular phenotypes that differ in potency. For instance, this strategy could be used to identify cells primed for differentiation within a multipotent or pluripotent cell population

normalized data matrix, because negative and 0 values are not allowed in SCENT. The reason why SCENT requires non-negatively valued data is that it estimates potency from a stochastic matrix on the network where the weights on the edges of the network represent signaling probabilities (which by necessity must be zero or positive). Usually, application of the *scater* pipeline will not result in negative values but will result in 0 values for those genes with zero counts. The reason why zero expression values must be further excluded is because the construction of the stochastic matrix involves ratios of gene expression values, and ratios will be undefined if the denominator is zero. Thus, for a count matrix $c_{ij}$ with rows labeling genes and columns labeling samples/cells, we may use a transformation of the form $\log_2\left\{c_{ij} \times \frac{sf}{tc_j} + 1.1\right\}$ where $tc_j$ is the total number of counts in cell $j$ and $sf$ is a global scale factor representing library complexity. The above transformation guarantees that values are always positive definite. Finally, we point out that another potential pitfall when preparing the scRNA-Seq data matrix for SCENT is to remove genes that are not expressed across all or a relatively high proportion of the cells (*see* also **Note 4**). Removing such nonexpressed genes is of course a very popular procedure in scRNA-Seq data analysis, but that is because such genes are uninformative or do not explain interesting variation across cells. However, in the context of estimating differentiation potency with SCENT, it is important to note that this estimation is done for each cell individually, i.e., independently of all other cells (*see* **Note 5**). If genes are truly not expressed in a given cell or in a given condition, or across a range of different conditions, then removing them could potentially lead to loss of information with regards to the signaling entropy calculation, as their removal may result in a much smaller network (*see* **Note 4**). Thus, for the specific task of estimating differentiation potency via SCENT, we do not recommend removing any genes based on zero or low expression levels across all or most cells.

*3.1.2 Integration with User-Defined Gene Functional Network*

Integration is achieved with the *DoIntegPPI* function, and consists of two steps:

1. The function takes as input two arguments, the normalized scRNA-Seq data matrix *exp.m* with rows labeling genes and columns labeling cells/samples, and a user-defined network *ppiA.m* with rows and columns labeling genes. The same gene identifier must be used for both expression and network matrices. The function finds the overlap between the gene identifiers specifying the network with those specifying the scRNA-Seq matrix, and then extracts the maximally connected subnetwork (*see* **Note 3**), specified by the *adjMC* output argument.

2. The function also constructs the reduced scRNA-Seq matrix, specified by the *expMC* output argument, and which is therefore matched to the rows (nodes) of the *adjMC* matrix.

*3.1.3 Computation of the Signaling Entropy Rate*

The output from *DoIntegPPI* is then used as input to the functions that compute the signaling entropy rate, denoted as *SR*. For a given fixed network, i.e., for the given *adjMC* matrix, there is a maximum possible *SR* value (denoted as *maxSR*), which is obtained for a particular edge-weight configuration [22]. It is thus very convenient to report the *SR* value of any given cell, normalized against this maximum possible value, which means that *SR* is then bounded between 0 and 1. The maximum entropy rate value can be calculated using the *CompMaxSR* function. This function takes as input the adjacency matrix, i.e., the *adjMC* output from the *DoIntegPPI* function, and returns the *maxSR* value as output.

Having obtained the normalization factor *maxSR*, we can then proceed to compute the *SR* value for any given cell/sample, using the *CompSRana* function. This function takes four objects as input:

1. The expression profile vector of the cell/sample (*exp.v*), which is a given column of the *expMC* output matrix from *DoIntegPPI*,

2. The network adjacency matrix (*adj.m*), i.e., the *adjMC* output from *DoIntegPPI*,

3. A logical parameter *local* to tell the function where to report back the local, i.e., gene-centric, signaling entropies, and,

4. *maxSR*, the maximum entropy rate calculated earlier. Specifying *maxSR = NULL* will force the function to return non-normalized *SR* values.

Two notes with the above procedure: (a) the local gene-based entropies can be used in downstream analyses for ranking genes according to differential entropy, but only if appropriately normalized. For instance, they could be used to identify the main genes driving changes in the global signaling entropy rate of the network. However, if the user only wishes to estimate potency, specifying *local = FALSE* is fine, which will save some RAM on the output object, (b) by specifying the input object as an expression vector, the user can easily use the *parallel* R-package to compute the *SR* values for all cells/samples simultaneously. For this purpose, we also provide on the github website another function *CompSRanaPRL*, which takes in as input an index and the full scRNA-Seq data matrix in place of the expression profile of one cell/sample. One can then use the *mclapply* function in the *parallel* package to loop over the index values, which specify the columns (i.e., cells/samples) for which the *SR* values are desired. This will become clearer in the example given further below. The output of the

*CompSRana* and *CompSRanaPRL* functions is a list consisting of four objects:

1. *sr*: the *SR* value of the cell/sample.

2. *inv*: a vector specifying the invariant measure, or steady-state probability, over the network. That is, this vector will be equal to the number of nodes in the adjMC matrix, and its entries must add to 1.

3. *s*: a vector containing the unnormalized local signaling entropies, and therefore a vector of length equal to the number of nodes in the adjMC matrix.

4. *ns*: if local = TRUE, a vector containing normalized local signaling entropies, and therefore a vector of length equal to the number of nodes in the adjMC matrix.

The computation of signaling entropy is relatively fast. After the integration of the scRNA-Seq data matrix with the network, the resulting graph has approximately 8000 nodes, and estimation of *SR* for one sample only takes 1–2 s on an Intel Xeon(R) CPU E3-1575 M v5 @ 3.00GHz processor.

### 3.2 Real Dataset Test

To illustrate the procedure above, we now run in detail through the example given in Table 1. As mentioned, we assume that the count matrix has been normalized, say using a specific package (e.g., *scater*). The normalized count matrix we use can be downloaded from *http://github.com/aet21/SCENT/scChu.Rd* and loaded into your session using the load command:

```
load("sceChu.Rd");
```

This loads in *sceChu.m*, the log-normalized count matrix, and *phenoSCchu.v*, an index vector specifying the cell-type (Table 1). Likewise, we need to load in the gene-functional (PPI) network specified earlier, using, e.g.,

```
load("hprdAsigH-13Jun12.Rd");
```

Thus, we would then run the following series of commands:

```
int.o <- DoIntegPPI(sceChu.m,hprdAsigH.m)
adjMC.m <- int.o$adjMC
maxSR <- CompMaxSR(adjMC.m)
idx.l <- as.list(1:ncol(scChu.m))
out.l <- mclapply(idx.l,CompSRanaPRL,int.o$expMC,adjMC.m,
maxSR,mc.cores=10);
```

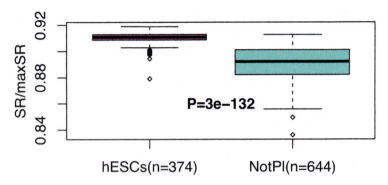

**Fig. 2** Boxplot of signaling entropy rate values (*SR*, *y*-axis) against cell-type (pluripotent hESCs vs non-pluripotent (NotPl), *x*-axis). *P*-value is from a one-tailed Wilcoxon rank sum test. Number of cells in each cell-type category is given in group labels

The last line assumes we are running on a computer with at least 10 processing cores. From the *out.l* object, we can then finally store the *SR* values in a vector which we shall denote by *SRpSC.v*.

To check that the *SR* values do indeed correlate with potency, we can run something like.

```
phenoChu.v <- phenoSCchu.v
phenoChu.v[phenoSCchu.v>=2] <- 2; ### to compare plurip. to
non-plurip.
boxplot(SRpSC.v ~ phenoChu.v)
```

which should result in the display shown in Fig. 2.

It might also be of interest to explore the heterogeneity in potency of the cell population. One can approach this question using the *DoSCENT* function, as shown below:

```
scent.o <- DoSCENT(sceChu.m,SRpSC.v,phenoSCchu.v)
```

The distribution of potency states in relation to cell-type can be obtained from the $distPSPH entry:

```
> scent.o$distPSPH
 ordpotS.v
celltype 1 2
1 355 19
2 90 83
3 8 61
4 0 159
5 0 105
6 10 128
```

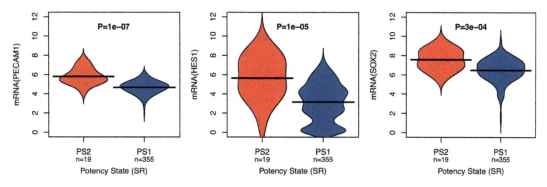

**Fig. 3** Normalized mRNA expression values for 3 genes that mark hESCs primed for differentiation into mesoderm (PECAM1) or ectoderm (SOX2 & HES1) lineages, against the predicted potency group of hESCs. PS1 = pluripotent/higher potency state, PS2 = non-pluripotent/lower potency state. *P*-values are from a one-tailed Wilcoxon rank sum test

This indicates that the algorithm inferred 2 potency states, with the first state being enriched for pluripotent hESCs (row indexed by 1) and a substantial proportion of neural progenitors (row indexed by 2), whereas the other potency state is dominated by non-pluripotent cell-types (row index values 2–6). Of note, approximately 5% of the hESCs were predicted to be of significantly lower potency than the rest, and it is natural to assume that these cells may already be primed to differentiate into any one of the different germ layers (ectoderm, mesoderm, endoderm). We can check that the predicted lower potency for these hESCs is in line with this hypothesis, by comparing the expression value distribution of a known mesoderm stem-cell marker (PECAM1) and two neural stem cell markers (HES1 and SOX2), all of which exhibited higher expression in hESCs categorized into the lower potency group (Fig. 3), as required.

# 4   Notes

1. Signaling entropy has been validated as a measure of potency across a large number of independent datasets, including time-course and non-time-course scRNA-Seq sets [22], as well as time-course and non-time-course bulk RNA-Seq and microarray sets [24]. Its robustness stems from three properties: First, it integrates the expression data with orthogonal systems-level information, as that provided by a PPI network. Although these networks are themselves noisy, *SR* depends mainly on the relative connectivity of the proteins in the network, which is a robust feature. The connectivity is important since *SR* can be approximated as a suitably transformed correlation between

the transcriptome and connectome, with cells of increased potency exhibiting higher mRNA expression levels of the more connected or promiscuous proteins [22]. Second, expression values only enter the $SR$ calculation as expression ratios, therefore rendering the $SR$ value fairly insensitive to the absolute scale and normalization procedure for the expression data (although as mentioned earlier, a log-transformation is recommended to regularize the effects of outlier expression). Third, $SR$ is a global measure computed over a large network of over 8000 nodes. In the case of scRNA-Seq data, this confers it substantial robustness against technical dropouts. Moreover, technical dropouts are more likely to affect lowly expressed genes, which, theoretically, for the networks considered here, are less influential in determining the $SR$ value.

2. We emphasize that the use of a PPI network is key. As shown by us previously, randomizing the expression values over the network, which in effect randomly reassigns connectivity values to each gene, results in a substantially reduced discrimination [22]. Likewise, we find that reverse-engineering a network from the data itself does not yield a robust potency measure. As mentioned in the previous note, potency is encoded in part via a subtle positive correlation between the transcriptome and PPI connectome. Thus, we strongly recommend a PPI network such as those derived from Pathway Commons (https://www.pathwaycommons.org/).

3. It is important that the signaling entropy rate is estimated over a connected network, since the entropy rate is ill-defined for networks that are not. The function within the package that integrates the scRNA-Seq data with the user-specified PPI network will automatically identify the maximally connected subnetwork, to ensure that the network passed on to the subsequent functions that compute $SR$ is connected. It is important however to check the size of the maximally connected subnetwork before running the final computation of $SR$. Usually, the maximally connected subnetwork will be a "prominent" subnetwork, i.e., one containing a large fraction (say over 80% or 70%) of the original number of nodes in the user-specified PPI network. In our experience, it is very unlikely that, for instance, upon integration with the scRNA-Seq data, the original PPI network would split up into 2 large and roughly equally sized components.

4. It is important to point out again that the robustness of Signaling Entropy derives in part from using a large (connected) network. It is therefore important not to remove genes from the scRNA-Seq data matrix before integration with the PPI

network, as removing a large number of such genes prior to integration could result in a network that is too small, or may not result in a sufficiently large maximally connected subnetwork. By sufficiently large we mean a network of at least 5000 nodes or so. A common procedure in scRNA-Seq data analyses is however to remove genes that are not expressed across all or a great proportion of the cells, since these are naturally uninformative for popular tasks such as dimensional reduction, clustering, or differential expression. However, the entropy calculation for estimating potency is different in that it estimates it for each cell independently of all others, and so removing genes that are not expressed could potentially remove biological information by unnecessarily reducing the size of the network, and thus compromising the accuracy of *SR*.

5. We stress that Signaling Entropy represents a model-driven general approach to the estimation of differentiation potency. It is estimated for each cell or sample individually without using information from other cells or samples. As such, it is relatively assumption-free, does not require training or feature selection, thus avoids overfitting, and in principle allows ordering of cells in terms of their potency within any given lineage. However, we point that the full resolution of the method, i.e., how big changes in potency can be detected, is still under active investigation.

# 5 Conclusions

Computation of signaling entropy can help users order cells/samples according to their potency, without the need for any prior biological knowledge, input, or feature selection. This may help identify in an unbiased way novel stem-and-progenitor cell phenotypes in large-scale scRNA-Seq data, specially non-time-course data, and in both normal and cancer physiological settings. The functions provided in the SCENT package and the example workflow provided here should help users learn how to compute signaling entropy for their single-cell or bulk RNA-Seq data.

## References

1. Waddington CH (1966) Principles of development and differentiation. Macmillan, London, pp 1905–1975

2. Moris N, Pina C, Arias AM (2016) Transition states and cell fate decisions in epigenetic landscapes. Nat Rev Genet 17:693–703. https://doi.org/10.1038/nrg.2016.98

3. Levsky JM (2002) Single-cell gene expression profiling. Science 297:836–840. https://doi.org/10.1126/science.1072241

4. Laurenti E, Göttgens B (2018) From haematopoietic stem cells to complex differentiation landscapes. Nature 553:418–426. https://doi.org/10.1038/nature25022

5. Lang AH, Li H, Collins JJ, Mehta P (2014) Epigenetic landscapes explain partially reprogrammed cells and identify key reprogramming genes. PLoS Comput Biol 10:e1003734. https://doi.org/10.1371/journal.pcbi.1003734

6. Tirosh I, Venteicher AS, Hebert C et al (2016) Single-cell RNA-seq supports a developmental hierarchy in human oligodendroglioma. Nature 539:309–313. https://doi.org/10.1038/nature20123

7. Tirosh I, Izar B, Prakadan SM et al (2016) Dissecting the multicellular ecosystem of metastatic melanoma by single-cell RNA-seq. Science 352:189–196. https://doi.org/10.1126/science.aad0501

8. Wang Z, Gerstein M, Snyder M (2009) RNA-Seq: a revolutionary tool for transcriptomics. Nat Rev Genet 10:57–63. https://doi.org/10.1038/nrg2484

9. Grün D, van Oudenaarden A (2015) Design and analysis of single-cell sequencing experiments. Cell 163:799–810. https://doi.org/10.1016/j.cell.2015.10.039

10. Trapnell C, Cacchiarelli D, Grimsby J et al (2014) The dynamics and regulators of cell fate decisions are revealed by pseudotemporal ordering of single cells. Nat Biotechnol 32:381–386. https://doi.org/10.1038/nbt.2859

11. Marco E, Karp RL, Guo G et al (2014) Bifurcation analysis of single-cell gene expression data reveals epigenetic landscape. Proc Natl Acad Sci U S A 111:E5643–E5650. https://doi.org/10.1073/pnas.1408993111

12. Setty M, Tadmor MD, Reich-Zeliger S et al (2016) Wishbone identifies bifurcating developmental trajectories from single-cell data. Nat Biotechnol 34:637–645. https://doi.org/10.1038/nbt.3569

13. Bendall SC, Davis KL, E-AD A et al (2014) Single-cell trajectory detection uncovers progression and regulatory coordination in human B cell development. Cell 157:714–725. https://doi.org/10.1016/j.cell.2014.04.005

14. Chen J, Schlitzer A, Chakarov S et al (2016) Mpath maps multi-branching single-cell trajectories revealing progenitor cell progression during development. Nat Commun 7:11988. https://doi.org/10.1038/ncomms11988

15. Qiu X, Mao Q, Tang Y et al (2017) Reversed graph embedding resolves complex single-cell trajectories. Nat Methods 14:979–982. https://doi.org/10.1038/nmeth.4402

16. Rizvi AH, Camara PG, Kandror EK et al (2017) Single-cell topological RNA-seq analysis reveals insights into cellular differentiation and development. Nat Biotechnol 35:551–560. https://doi.org/10.1038/nbt.3854

17. Haghverdi L, Büttner M, Wolf FA et al (2016) Diffusion pseudotime robustly reconstructs lineage branching. Nat Methods 13:845–848. https://doi.org/10.1038/nmeth.3971

18. Angerer P, Haghverdi L, Büttner M et al (2016) Destiny: diffusion maps for large-scale single-cell data in R. Bioinformatics 32:1241–1243. https://doi.org/10.1093/bioinformatics/btv715

19. Chu L-F, Leng N, Zhang J et al (2016) Single-cell RNA-seq reveals novel regulators of human embryonic stem cell differentiation to definitive endoderm. Genome Biol 17:2315. https://doi.org/10.1186/s13059-016-1033-x

20. Grün D, Muraro MJ, Boisset J-C et al (2016) De novo prediction of stem cell identity using single-cell Transcriptome data. Cell Stem Cell 19:266–277. https://doi.org/10.1016/j.stem.2016.05.010

21. Guo M, Bao EL, Wagner M et al (2017) SLICE: determining cell differentiation and lineage based on single cell entropy. Nucleic Acids Res 45:e54. https://doi.org/10.1093/nar/gkw1278

22. Teschendorff AE, Enver T (2017) Single-cell entropy for accurate estimation of differentiation potency from a cell's transcriptome. Nat Commun 8:15599. https://doi.org/10.1038/ncomms15599

23. Gómez-Gardeñes J, Latora V (2008) Entropy rate of diffusion processes on complex networks. Phys Rev E Stat Nonlinear Soft Matter Phys 78:114. https://doi.org/10.1103/PhysRevE.78.065102

24. Banerji CRS, Miranda-Saavedra D, Severini S et al (2013) Cellular network entropy as the energy potential in Waddington's differentiation landscape. Sci Rep 3:1129. https://doi.org/10.1038/srep03039

25. Teschendorff AE, Sollich P, Kuehn R (2014) Signalling entropy: a novel network-theoretical framework for systems analysis and interpretation of functional omic data. Methods 67:282–293. https://doi.org/10.1016/j.ymeth.2014.03.013

26. Banerji CRS, Severini S, Caldas C, Teschendorff AE (2015) Intra-tumour Signalling entropy determines clinical outcome in breast and lung cancer. PLoS Comput Biol 11:e1004115. https://doi.org/10.1371/journal.pcbi.1004115

27. Lun ATL, McCarthy DJ, Marioni JC (2016) A step-by-step workflow for low-level analysis of single-cell RNA-seq data with bioconductor. F1000Res 5:2122. https://doi.org/10.12688/f1000research.9501.2

28. McCarthy DJ, Campbell KR, Lun ATL, Wills QF (2017) Scater: pre-processing, quality control, normalization and visualization of single-cell RNA-seq data in R. Bioinformatics 247:btw777. https://doi.org/10.1093/bioinformatics/btw777

29. Butler A, Hoffman P, Smibert P et al (2018) Integrating single-cell transcriptomic data across different conditions, technologies, and species. Nat Biotechnol 36:411–420. https://doi.org/10.1038/nbt.4096

# Chapter 10

## Inference of Gene Co-expression Networks from Single-Cell RNA-Sequencing Data

### Alicia T. Lamere and Jun Li

### Abstract

Single-cell RNA-Sequencing is a pioneering extension of bulk-based RNA-Sequencing technology. The "guilt-by-association" heuristic has led to the use of gene co-expression networks to identify genes that are believed to be associated with a common cellular function. Many methods that were developed for bulk-based RNA-Sequencing data can continue to be applied to single-cell data, and several of the most widely used methods are explored. Several methods for leveraging the novel time information contained in single-cell data when constructing gene co-expression networks, which allows for the incorporation of directed associations, are also discussed.

**Key words** Gene co-expression network, Gene regulatory network, Single-cell RNA-Seq, Correlation coefficient, Count data, Directed network, Pseudotime

## 1 Introduction

Both gene function and regulatory relationships can be inferred and identified through the use of gene co-expression networks (GCNs). In GCNs, nodes represent genes and an edge between two nodes represents co-expression between a pair of genes (*see* Fig. 1a below). Computational inference of GCNs is based on a set of experiments, each measuring the expression of a large set of genes. These experiments use samples from different tissues or different conditions, so genes that are co-expressed tend to have high/low expressions in the same experiments simultaneously.

The "guilt-by-association" heuristic has led to the use of GCNs to identify co-expressing genes that are believed to be associated with a common cellular function. Well-constructed GCNs are used to help understand molecular mechanisms underlying biological processes and to predict gene functions that are not previously known [1].

Many methods have been proposed for constructing GCNs based on microarray and bulk-based RNA-Sequencing (RNA-Seq)

Guo-Cheng Yuan (ed.), *Computational Methods for Single-Cell Data Analysis*, Methods in Molecular Biology, vol. 1935, https://doi.org/10.1007/978-1-4939-9057-3_10, © Springer Science+Business Media, LLC, part of Springer Nature 2019

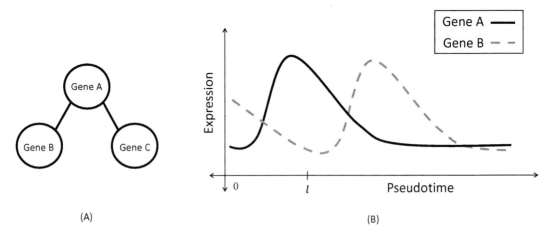

**Fig. 1** (**a**) Example GCN. Here, because there exists an edge between the nodes representing Gene A and Gene B, these two genes show evidence of co-expression. Meanwhile, no edge exists between Gene B and Gene C, so there is no evidence of co-expression for this pair of genes. (**b**) Example of two gene expressions that exhibit a regulatory relationship when ordered by pseudotime. If we simply considered their correlation, the pair appear unrelated. However, if we correlate the lagged expression, looking at Gene A's expression from time 0 and Gene B's expression from time *l*, then they exhibit a strong positive correlation

data [2–6] and can be applied to single-cell RNA-Sequencing (scRNA-Seq) to construct GCNs as well. A significant portion of this chapter will be dedicated to discussing these methods, with particular focus on so-called "correlation-based" methods, which remain the most straightforward and easy-to-interpret methods for constructing GCNs. In these methods, an edge exists between two genes if a strong correlation is found between the expression of that pair of genes.

However, in using these GCNs with scRNA-Seq, the data are treated no differently than bulk-based RNA-Seq. As a result, information contained uniquely in scRNA-Seq is lost. Particularly, correlation-based GCNs constructed based on microarray or RNA-Seq data only capture simultaneous associations between pairs of genes. Biologically, if a gene enhances/inhibits another gene, then the latter gene will have delayed expression/silence [7]. For such a pair, the co-expression of the two genes is strong if the delay in time is taken into account, but can be weak if only simultaneous association is explored.

Consider the hypothetical gene expressions in Fig. 1b, arranged by time. If we consider the entire expressions of genes A and B, then their correlation is close to 0. However, looking at their expressions, it clearly appears that gene A is enhancing the expression of gene B—meaning that as gene A's expression increases, we see a delayed decrease in gene B's expression, beginning at time $t = l$. We also observe a corresponding delayed decrease in gene B's expression following a decrease in gene A. This implies a strong positive correlation, which we can capture if we take our time lag into account.

A pioneering extension of the bulk-based RNA-Seq technique, scRNA-Seq, is able to capture this time information in an indirect way. scRNA-Seq measures the gene expression profile of each individual cell, with hundreds to thousands of cells in a single run. These cells are at different time points of their cell cycles, and these time points can be estimated based on the idea that expression profiles should be similar in cells at similar time points. These estimated time points are called "pseudotime," and several algorithms have been developed for their estimation [8–12]. We also discuss these methods in more detail below.

## 2     Materials

1. R software for statistical computing

   R Packages:
2. edgeR (available through Bioconductor).
3. WGCNA (available through CRAN).
4. DiPhiSeq (available through CRAN).
5. Monocle (available through Bioconductor).
6. LEAP (available through CRAN).

## 3     Methods

This section summarizes the basic steps involved when constructing a GCN from single-cell data, while exploring several methods that can be used for each step.

### 3.1     Normalization

Normalization is an important question to consider when working with scRNA-Seq data. Beyond differences in sequencing depth that must be taken into account between experiments, as with bulk-based RNA-Seq, scRNA-Seq experiments also tend to have a large amount of technical noise (*see* **Note 1**). While there do exist GCN construction methods that do not require normalization, none are correlation-based. Commonly used bulk-based normalization methods such as FPKM [13], upper quartile [14], or trimmed mean of M-values [15] can be applied to scRNA-Seq (*see* **Note 2**). These normalizations can be easily performed through the use of the edgeR package [16] using the following steps:

1. Let data be a matrix of raw read counts.
2. Find the normalization factors for the method you would like to use. For example, if we wish to use trimmed mean of M-values:

```
>norm_factors = calcNormFactors(data, method="TMM")
```

3. Output a matrix of normalized counts by using the cpm function:

```
>norm_data = cpm(norm_factors,
normalized.lib.sizes = TRUE)
```

**3.2  Identifying Highly Expressed Genes**

After normalization and before constructing any GCNs, it usually is important to filter your dataset to only those showing reasonably high average expression and range of expression. This step is particularly important for scRNA-Seq data which has a large number of drop-out events. By filtering to keep genes with not only moderate/high expression, but a large range of expression, we can be more confident that the edges identified in our network are not simply the result of noise in the dataset. One method for identifying these genes is to:

Find the average and inter-quartile range of the normalized expression values for each gene. In R, this can easily be done using base functions:

```
>iqr_vals = apply(norm_data, 1, IQR)
>mean_val = apply(norm_data, 1, mean)
```

1. Scale the interquartile range for each gene by dividing by its average expression:

```
>iqr_scale = iqr_vals/mean_val
```

2. Select genes to keep for analysis with a sufficiently high average expression and scaled interquartile range. What qualifies as sufficient is up to the user's discretion, however, a scaled IQR of at least 0.5 and a mean of at least 0.1 are usually desirable (*see* **Note 3**). The filtered dataset can be obtained by using the code:

```
>keep = (mean_val >= 0.1 & iqr_scale >= 0.5)
>filter_data = norm_data[keep,]
```

**3.3  Methods for Constructing Traditional GCNs**

When constructing traditional gene co-expression networks, meaning they are undirected, a variety of options exist already for use with bulk-based RNA-Seq data. These methods can be broken into two key categories: correlation-based and non-correlation-based. Correlation-based methods are generally faster and simpler to implement, and hence are often preferred, while non-correlation based are more complicated and computationally intensive and hence will not be discussed in detail here (*see* **Note 4**). Conceptually, all of these methods may still be applied to scRNA-Seq data.

Generally, correlation-based methods take the matrix of expression data and calculate the correlation for each gene pair, usually through the calculation of Pearson's correlation coefficient. These methods remain the most straightforward and easy to understand methods for constructing GCNs for RNA-Seq data.

### 3.3.1  WGCNA

Weighted Gene Co-expression Network Analysis (WGCNA) was originally developed for the analysis of microarray data, which is continuous [17]. Hence, to use WGCNA, the count data must first be normalized using the methods described above. Then, the Pearson's correlation coefficient is calculated for each pair of genes, resulting in an adjacency matrix to which a hard or soft threshold can then be applied to determine a particular GCN. This method has been successfully applied to RNA-Seq data after proper normalization [18–20]. One advantage of this method is its use of a soft-threshold through the incorporation of weighted edges. This allows the user to rank edges for consideration based on the strength of the connection between each pair of genes. To implement WGCNA, the following code is provided through a Tutorial written by the authors:

1. To speed up calculations, enable the use of multi-threading:

```
>enableWGCNAThreads()
```

2. Identify the soft-threshold to be used:

```
>Powers = c(c(1:10), seq(from=12, to=20, by=2)
>soft = pickSoftThreshold(filter_data, powerVectors =
power, verbose = 5)
>R_sqr= -sign(soft$fitIndices[,3]) *soft$fitIndices[,2]
>plot(soft$fitIndices[,1], R_square)
```

3. Based on this plot, choose the lowest power for which the signed R square curve flattens out upon reaching a high value. Let us say that this value is 6, we can then construct our network with the code:

```
>network = blockwiseModules(filter_data, power = 6, TOMType
= "unsigned", minModuleSize = 30, reassignThreshold = 0,
mergeCutHeight = 0.25, numericLabels = TRUE,
pamRespectsDendro = FALSE, saveTOMs = TRUE, saveTOMFileBase
= "My_network" verbose = 3)
```

The network will be contained in the TOM file saved as "My_network".

### 3.4  iCC

Distribution-inversed and Gaussian-transformed Correlation Coefficient is a GCN construction method developed directly for use on RNA-Seq data [21]. This method does not require any kind of normalization beforehand. Instead, it works with the count data directly and incorporates the sequencing depth into the model thereby increasing the power. Though not provided as a package, this method is relatively simple to implement by hand:

1. Let `filter_data` be data that have not been normalized, but have been filtered to keep the most highly expressed genes. Find the transformed sequencing depth for each experiment:

```
>d = colMeans(filter_data)
>depth = exp(log(d) - mean(log(d)))
```

2. Estimate the parameters for the distributions describing each gene's expression across each experimental condition. This can be done with the DiPhiSeq R package [22] using the following code for genes $i$ and $j$:

```
>results_i = robnb(filter_data[i,], depth)
>results_j = robnb(filter_data[j,], depth)
```

3. Calculate the probability of observing each expression count k using the distribution parameters:

```
>for (k in (1: length(depth)){
> pval_i[k] = pnbinom(filter_data[j,k],
>size=1/results_i$phi, mu=exp(results_i$beta)*depth[k])
>pval_j[k] =pnbinom(filter_data[j,k], size=1/results_j$phi,
mu=exp(results_j$beta)*depth[k])}
```

4. Use these probabilities to find the corresponding standard-Normal distribution values.

```
>norm_i = qnorm(pval_i, 0, 1)
>norm_j = qnorm(pval_j, 0, 1)
```

5. For each gene pair, use the now Gaussian-distributed values to estimate their correlation.

```
>corr_ij = cor(norm_i, norm_j)
```

These correlations define the adjacency matrix that describes the GCN. Typically, a cutoff should be chosen for the correlations to construct the network. It is generally recommended that absolute values of 0.7 or more be used, while not going below 0.5 as these edges have a greater chance of being the result of noise in the data.

*3.4.1 Directed GCNs Through Pseudotime for scRNA-Seq Data*

The network construction methods explored above only describe co-expression of genes. They were originally designed for use on either microarray or bulk-based RNA-Seq data, and therefore do not leverage any of the additional information available through scRNA-Seq. As described in the introduction, we can leverage cell-cycle time information through pseudotime algorithms to identify directed relationships between genes and construct directed networks. Here we discuss one method, LEAP, in particular, but there are many other methods available for constructing GCNs by leveraging the information contained uniquely in scRNA-Seq data (*see* **Note 5**).

3.4.2  *Estimating Pseudotime*

Key to any time-based inference on gene expression data is the method used to estimate the time information. For scRNA-Seq, several algorithms have been developed for the estimation of these time points, called "pseudotime." In general, the pseudotime analysis of the scRNA-Seq data often involves dimension reduction methods to deal with high-dimensionality, due to the often thousands of gene expression levels measured in each sample. There are many pseudotime algorithms available now (*see* **Note 6**). We will use the method Monocle because of its ease of use and implementation as an R package. The authors of Monocle [8] recognize that by projecting the data to a lower dimensional space, natural clustering of cells can occur and this clustering can capture cells at different time points. Their algorithm works by first representing each cell's expression as a point in a Euclidean space with dimensions representing each gene included in the sample. Then this high-dimensional space is reduced using independent component analysis (ICA), which, as its name implies, projects the gene expression profiles into a lower-dimensional space that best distinguished the independent components—or in our case, cells. The algorithm then constructs a minimum spanning tree on the cells in this lower-dimensional space. This tree is simply the shortest path that connects all cells without revisiting any edges. Finally, cells' positions in the minimum spanning tree are used to assign "pseudotime" values (*see* **Note 7**). Monocle does not require the scRNA-Seq data to be normalized beforehand. Instead, it handles all normalization internally. As the first step in our network construction, we will use Monocle to generate the pseudotime for each cell in the scRNA-Seq dataset using genes that are known to be associated with cell cycle.

1. Let `filter_data` be data that have not been normalized, but have been filtered to keep the most highly expressed genes. Create a data set object for monocle to use:

```
>mon_data= newCellDataSet(as.matrix(filter_data))
```

2. Reduce the dimensions:

```
>red_data = reduceDimension(mon_data)
```

3. Calculate the pseudotime orderings:

```
>ord_data = orderCells(red_data, num_path = 2, reverse =
FALSE)
>pseudotime = pData(ord_data)
```

*3.4.3  LEAP Algorithm for Estimating Co-expression*

LEAP is a method created for direct use on scRNA-Seq data to construct directed GCNs. Borrowing from time-series analysis, LEAP sorts cells according to the estimated cell-cycle-based pseudotime creating a "pseudotime-series," and then computes the maximum correlation over all possible time lags [23]. This maximum correlation is used as the statistic to replace the traditional Pearson's correlation coefficient for constructing the gene co-expression network, and the statistical significance of this statistic is measured by the false discovery rate (FDR) calculated using permutations. LEAP is implemented as an R package, and the general steps for generating a GCN are:

1. Sort the cells in your scRNA-Seq dataset according to the pseudotime you calculated using Monocle.

2. In order to apply LEAP, the scRNA-Seq expression counts must be normalized using any of the methods described above. Usually, TMM is recommended.

3. Apply the MAC_counter() function to your normalized dataset to generate the correlation matrix that the GCN will be based on. By maximizing over all possible time lags, the correlations found by LEAP can often be larger than a traditional Pearson's correlation coefficient (*see* **Note 8**). The following is the R code to estimate the directed GCN:

```
>MAC_results = MAC_counter(data=filter_data,
max_lag_prop=1/3, MAC_cutoff=0.2, lag_matrix=T)
```

It is important to notice that, in general, the gene pairs *(i,j)* and *(j,i)* will not have the same maximum absolute correlation. This is because for the pair *(i,j)*, the maximum absolute correlation that is found measures the co-expression when gene *j*'s expression is delayed compared to the expression of gene *i*. As a result, LEAP is able to capture directional relationships, and these directional relationships likely imply regulatory relationships. Therefore, LEAP is able to extend the information contained within traditional GCNs—Incorporating enhancing, inhibiting, and co-expression relationships between genes into one single network.

4. Next we need to determine direction of correlations. Note that by considering the sign of the maximum correlation and the lag at which it occurs, the direction of the regulatory relationship between a pair of genes is determined. For example, a positive correlation between a pair A and B occurring at a nonzero lag suggests that gene A enhances gene B, meaning that an increase in gene A is causing an increase in gene B's expression. Conversely, a negative correlation between a pair A and B occurring at a nonzero lag suggests that an increase in gene A is causing a decrease in gene B's expression. Note that if these maximum correlations occurred when the pair was considered gene B and gene A, then B would be up/downregulating gene A. Finally, if the maximum correlation occurs when the lag is 0, then a co-expression relationship has been captured (as is identified by traditional correlation-based GCNs), and suggests that gene A and gene B are both regulated by a third gene. These lag results are generated by the MAC_counter function and saved as a "lag" file.

5. Finally, LEAP incorporates a method to determine the appropriate cutoff to use to construct the network while controlling false discoveries. To estimate the false discovery rate (FDR), each gene $i$'s normalized expression counts are permuted a default number of 100 times. Then for each permutation, an estimated maximum absolute correlation is calculated. For each correlation cutoff, an FDR is then estimated. These estimated FDRs can then be used to determine the appropriate correlation cutoff that should be used to control the false discoveries at the desired rate.

```
>MAC_perm(data=filter_data,MACs_observ=MAC_results, num_perms=10, max_lag_prop=1/3,
FDR_cutoffs=101)
```

This cutoff can then be applied to the maximum absolute correlation matrix to create a directed GCN for the scRNA-Seq data.

# 4  Notes

1. *ERCC Spike-Ins.* It has been observed that the use of normalization methods that incorporate spike-in External RNA Control Consortium (ERCC) levels tends to provide better removal of the frequently high technical noise found in scRNA-Seq [24]. If ERCCs have been incorporated in your scRNA-Seq data, then methods such as removed unwanted variants [25] and gamma regression model [24] should be used.

2. *Log-transformation*. It can often be helpful to apply an additional $\log(x + 1)$ transformation to expression data after normalization to reduce the effects of particularly large expression values.

3. *Choosing cutoff for identifying highly expressed genes*. The choice of cutoffs for average expression and scaled interquartile range should be determined based on each particular scRNA-Seq dataset, and in practice must often be tweaked to identify the most informative and practical set of genes. It is also important for most methods to reduce the number of genes to a size that is computationally feasible. In practice, one-thousand genes or less works well for most methods on a typical laptop computer.

4. *Non-correlation-based GCN construction methods*. There exist many methods for constructing GCNs from RNA-Seq data directly that do not involve estimating the correlation between genes. Instead, they use more sophisticated methods such as Markov Random Fields, partial-correlation, and mutual information. A downside to these methods is that they tend to require much more computational power and time to run on the large datasets common to RNA-Seq data, often working effectively on a set of only a few dozen to a few hundred genes. Three of the most widely used non-correlation-based methods are XMRF, GeneNet, and ARACNE. XMRF is an R package that includes four Markov Random Field, or Markov Network-based methods for constructing GCNs [26]. The methods are: a regular Poisson graphical model, which is only able to capture negative conditional dependencies between genes; a truncated Poisson graphical model that reduces the effects of large counts while allowing for positive and negative conditional dependencies; a sublinear Poisson graphical model that softens the reductions of the large counts; and finally, a local Poisson graphical model that, through the use of approximation, is in practice much faster [27]. Consequently, the package authors generally recommend the use of the local Poisson graphical model (LPGM) for RNA-Seq data. GeneNet is an R package that constructs "causal" or directed networks through the converting of a correlation network into a partial correlation network [28]. By doing so, these networks have the ability to essentially measure the relationships between genes after removing the effects of other genes on their correlation. This method was originally designed for use on microarray data, so once again RNA-Seq data must be normalized using the methods described above before GeneNet can be used to construct a GCN. One potential concern for GeneNet is that to perform the transformation from correlation matrix to partial-correlation matrix, the original correlation matrix must be positive definite. Usually, the large size of RNA-Seq datasets

resolves this issue as long as the number of genes in the dataset is much larger than the number of experiments. The third method, Algorithm for the Reconstruction of Accurate Cellular Networks (ARACNE), is information-theory based [29]. It has been shown that by using mutual information, a more accurate depiction of network relationships can be found when working with microarray data. ARACNE uses mutual information to determine dependencies among gene—Essentially measuring how much knowing the expression of one gene reduces the uncertainty of the expression of the other gene. A benefit of using mutual information to construct GCNs is that there is no monotonic relationship assumption, allowing them to capture nonlinear dependencies missed by correlation-based networks. This method was also originally designed for use on microarray data, so once again RNA-Seq data must be normalized before applying ARACNE. Another potential downside to the use of this algorithm is the longer computational time.

5. *Other directed construction methods.* Ocone et al. [30] have described a similar framework to LEAP using pseudotime that instead uses ordinary differential equations (ODE)-based methods to construct GCNs directly from scRNA-Seq data. Although not available as an R package, Matlab and C++ code are available for implementing it through the authors. The framework begins with dimension reduction. Ocone et al. recommend using the nonlinear method diffusion map [31] for dimension reduction prior to estimating pseudotime. After dimension reduction, an ad hoc clustering method is applied to separate branches associated with different cellular processes. This clustering depends on the user's choice of an initial cell, which is generally sufficiently identified by examining the visual layout of the diffusion map. Wanderlust is then applied to each branch identified through clustering. It is important to note that Wanderlust should be applied to the original expression data, not the dimension-reduced data. As a guide for the ODE model selection, the authors recommend first estimating an approximate network structure by applying GENIE3 [32]. This is then used as a starting place for the ODE model estimation, thereby reducing the number of models that would need to be compared. Finally, through MCMC iterations, the parameters for each ODE model are estimated using the pseudotime ordering of the cells in each branch. The best model is selected by comparing AIC, BIC, and Bayes' factors. Similarly, an alternative method called PIDC is mutual information-based using partial information decomposition, and can be combined with clustering methods or pseudotime-orderings to infer GCNs for scRNA-Seq data [33]. This method, through its use of MI, is able to capture nonlinear relationships

between pairs and groups of genes. The first three steps for the ODE framework described above can be implemented to discern a computationally feasible subset of genes to work with. Although not available as an R package, Julia code implementing PIDC is available through the authors. A third method is designed for experiments collected at specific time points. These time points may also contain important information for constructing GCNs. The algorithm SINCERITIES [34] uses ridge regression and partial correlation analysis to directly incorporate temporal changes in expression that are observed through these time points. Though not implemented as an R package, the R and Matlab code are both available through the authors.

6. *Wanderlust*. Another popular pseudotime estimation method is Wanderlust. Instead of using ICA, Wanderlust takes the high-dimensional data and transforms it into a nearest-neighbor graph—Meaning cells with similar expression profiles will be connected [12]. It then repeatedly identifies the shortest path and takes the average, using a cell's placement along this average path to determine its pseudotime.

7. *Estimating Pseudotime with Monocle*: Monocle's demonstrated effectiveness and ease of implementation tend to make it easier implement. However, when using Monocle it is important to pay attention to the "states" in which cells are classified and note that pseudotimes from one state do not correspond to the times in other states. As a result, each state should be treated separately. In practice, there usually is a state that most cells are captured by and hence analysis may be restricted to those cells.

8. *Choosing window size for LEAP*: Note that the default for this function is a maximum window size of two-thirds of the number of cells present in the dataset. Deviating significantly from this size may result in correlations that are artifacts of noise in the expression profiles rather than capturing true biological effects.

## References

1. Wolfe C, Kohane I, Butte A (2005) Systematic survey reveals general applicability of "guilt-by-association" within gene coexpression networks. BMC Bioinformatics 6(1):227

2. Stuart JM, Segal E, Koller D, Kim SK (2003) A gene-coexpression network for global discovery of conserved genetic modules. Science 302 (5643):249–255

3. Schafer J, Strimmer K (2005) An empirical bayes approach to inferring large-scale gene association networks. Bioinformatics 21 (6):754–764

4. Lee HK et al (2004) Coexpression analysis of human genes across many microarray data sets. Genome Res 14(6):1085–1094

5. Persson H et al (2005) Identification of genes required for cellulose synthesis by regression analysis of public microarray data sets. Proc Natl Acad Sci U S A 102(24):8633–8638

6. Basso K et al (2005) Reverse engineering of regulatory networks in human b cells. Nat Genet 37(4):382–390

7. Munksy B, Neuert G, van Oudenaarden A (2012) Using gene expression noise to understand gene regulation. Science 336 (6078):183–187

8. Trapnell C et al (2014) The dynamics and regulators of cell fate decisions are revealed by pseudotemporal ordering of single cells. Nat Biotechnol 32(4):381–386

9. Campbell K, Yau C (2015) Bayesian Gaussian process latent variable models for pseudotime inference in single-cell rna-seq data. bioRxiv. p. 026872

10. Reid JE, Wernisch L (2016) Pseudotime estimation: deconfounding single cell time series. Bioinformatics 32(19):2973–2980

11. Campbell K, Ponting CP, Webber C (2015) Laplacian eigenmaps and principal curves for high resolution pseudotemporal ordering of single-cell rna-seq profiles. bioRxiv. p 027219

12. Bendall SC et al (2014) Single-cell trajectory detection uncovers progression and regulatory coordination in human b cell development. Cell 157(3):714–725

13. Garber M et al (2011) Computational methods for transcriptome annotation and quantification using rna-seq. Nat Methods 8(6):469

14. Bullard JH et al (2010) Evaluation of statistical methods for normalization and differential expression in mrna-seq experiments. BMC Bioinformatics 11(1):94

15. Robinson MD, Oshlack A (2010) A scaling normalization method for differential expression analysis of rna-seq data. Genome Biol 11 (3):R25

16. Robinson MD, McCarthy DJ, Smyth GK (2010) edgeR: a bioconductor package for differential expression analysis of digital gene expression data. Bioinformatics 26:139–140

17. Langfelder P, Horvath S (2008) WGCNA: an R package for weighted correlation network analysis. BMC Bioinformatics 9(1):559

18. Iancu D et al (2012) Utilizing rna-seq data for de novo coexpression network inference. Bioinformatics 28(12):1592–1597

19. Kim H et al (2013) Peeling back the evolutionary layers of molecular mechanisms responsive to exercise-stress in the skeletal muscle of the racing horse. DNA Res 20(3):287–298

20. Xue Z et al (2013) Genetic programs in human and mouse early embryos revealed by single-cell RNA sequencing. Nature 500(7464):593

21. Specht AT, Li J (2015) Estimation of gene co-expression from rna-seq count data. Stat Interface 8(4):507–515

22. Li J, Lamere AT (2018) DiPhiSeq: Robust comparison of expression levels on RNA-Seq data with large sample sizes. Paper presented at the Joint Statistical Meetings, Vancouver, CA, 28 July–2 Aug 2018

23. Specht AT, Li J (2016) LEAP: constructing gene co-expression networks for single-cell rna-sequencing data using pseudotime ordering. Bioinformatics 33(5):764–766

24. Ding B, Zheng L, Wang W (2017) Assessment of single cell rna-seq normalization methods. G3 (Bethesda) 7(7):2039–2045

25. Risso D et al (2014) Normalization of rna-seq data using factor analysis of control genes or samples. Nat Biotechnol 32(9):896

26. Wan YW et al (2016) XMRF: an R package to fit markov networks to high-throughput genetics data. BMC Syst Biol 10(3):69

27. Allen GI, Liu Z (2013) A local poisson graphical model for inferring networks from sequencing data. IEEE Trans Nanobioscience 12 (3):189–198

28. Opgen-Rhein R, Strimmer K (2007) From correlation to causation networks: a simple approximate learning algorithm and its application to high-dimensional plant gene expression data. BMC Syst Biol 1(1):37

29. Margolin AA et al (2006) ARACNE: an algorithm for the reconstruction of gene regulatory networks in a mammalian cellular context. BMC Bioinformatics 7(1):S7

30. Ocone A et al (2015) Reconstructing gene regulatory dynamics from high-dimensional single-cell snapshot data. Bioinformatics 31 (12):i89–i96

31. Coifman RR et al (2005) Geometric diffusions as a tool for harmonic analysis and structure definition of data: diffusion maps. Proc Natl Acad Sci U S A 102(21):7426–7431

32. Huynh-Thu VA et al (2010) Inferring gene regulatory networks from expression data using tree-based methods. PLoS One 5(9): e12776

33. Chan TE et al (2017) Gene regulatory network inference from single-cell data using multivariate information measures. Cell Syst 5 (3):251–267

34. Papili Gao N et al (2017) SINCERITIES: inferring gene regulatory networks from time-stamped single cell transcriptional expression profiles. Bioinformatics 34(2):258–266

# Chapter 11

# Single-Cell Allele-Specific Gene Expression Analysis

## Meichen Dong and Yuchao Jiang

### Abstract

Allele-specific expression is traditionally studied by bulk RNA sequencing, which measures average gene expression across cells. Single-cell RNA sequencing (scRNA-seq) allows the comparison of expression distribution between the two alleles of a diploid organism, and characterization of allele-specific bursting. Here we describe SCALE, a bioinformatic and statistical framework for allele-specific gene expression analysis by scRNA-seq. SCALE estimates genome-wide bursting kinetics at the allelic level while accounting for technical bias and other complicating factors such as cell size. SCALE detects genes with significantly different bursting kinetics between the two alleles, as well as genes where the two alleles exhibit non-independent bursting processes. Here, we illustrate SCALE on a mouse blastocyst single-cell dataset with step-by-step demonstration from the upstream bioinformatic processing to the downstream biological interpretation of SCALE's output.

**Key words** Single-cell RNA sequencing, Allele-specific expression, Transcriptional bursting, Technical variability

## 1 Introduction

In diploid eukaryotic organisms, each autosomal gene has two copies/alleles: one inherited from the mother and one from the father [1–4]. Allele-specific expression (ASE) refers to the phenomenon that gene expression is unbalanced between the two alleles, and is found, in its extreme, in genomic imprinting, where the allele from one parent is uniformly silenced across cells, and in X-chromosome inactivation, where one of the two X-chromosomes in females is randomly silenced. During the past decade, ASE has been studied by bulk RNA sequencing (RNA-seq), where mean expression differences between the two alleles have been found across cells, in the form of allelic imbalance [5, 6]. Recent developments in single-cell RNA-seq (scRNA-seq) have made possible the better characterization of allelic differences in gene expression, which circumvent the averaging artifacts associated with traditional bulk population data [2, 4, 7, 8]. ASE analysis by scRNA-seq allows the comparison of expression distribution between the two alleles and characterization

Guo-Cheng Yuan (ed.), *Computational Methods for Single-Cell Data Analysis*, Methods in Molecular Biology, vol. 1935, https://doi.org/10.1007/978-1-4939-9057-3_11, © Springer Science+Business Media, LLC, part of Springer Nature 2019

of allele-specific transcriptional bursting, which is a fundamental property of gene expression where transcription from DNA to RNA occurs in bursts [7–10].

Here we describe SCALE (*Single-Cell ALlelic Expression*) [7], a systematic bioinformatic and statistical framework to study ASE in single cells by examining allele-specific transcriptional bursting kinetics. SCALE is comprised of three main steps. First, based on allele-specific read counts across cells, we adopt an empirical Bayes method to classify genes into three categories: silent, monoallelically expressed, or biallelically expressed. Next, for genes classified as biallelically expressed, we use a Poisson–Beta hierarchical model to estimate allele-specific bursting parameters while adjusting for dropout events, amplification and sequencing bias, and other complicating factors such as cell size. Finally, we apply a nonparametric bootstrap testing method to examine whether the two alleles of a gene share the same bursting parameters and perform a Chi-square test for independent bursting between the two alleles.

# 2   Materials

In this section, we outline the required materials for conducting ASE analysis by SCALE, including data input, computational environment, and software packages. In the Methods section that follows, we demonstrate how to carry out ASE analysis by scRNA-seq, from bioinformatic processing of the raw sequencing files to the application and result interpretation of SCALE. The bioinformatic pipeline and the R package for SCALE are available at https://github.com/yuchaojiang/SCALE.

## 2.1   Data Input

The data input for SCALE includes raw sequencing files from scRNA-seq studies (as fastq), as well as the corresponding genome assembly (as fasta). Based on the gene body coverage of the sequenced reads, scRNA-seq can be classified into two categories: full-transcript method (e.g., Smart-seq [11] and Smart-seq2 [12]) and tag method (e.g., Drop-seq [13] and the 10X Genomics Chromium Single Cell 3′ Solution [14]). To study ASE at germline heterozygous loci, full-transcript scRNA-seq protocol such as Smart-seq2 is recommended due to its broad coverage. To account for biases that are introduced during the library preparation and sequencing step, SCALE by default relies on external spike-ins, whose known concentration is used as ground truth for adjustment [15]. When spike-ins are not readily available, imputation-based methods such as SAVER [16] and scImpute [17] can be adopted to recover true underlying expression distribution.

## 2.2   Computational Environment

The upstream bioinformatic processing of raw sequencing data needs Linux or Unix platform in high-performance computing (HPC) system, which will return read count matrices as input for SCALE (*see* Subheading 3.1 for more details). The downstream ASE analysis can then be locally run on a Windows or Macintosh machine, where R is installed. For faster computation, SCALE can be run in parallel on an HPC cluster.

*2.3 Software Packages*

For read alignment, BWA [18] and STAR [19] are required for DNA and RNA sequencing, respectively. Picard Tools (http://broadinstitute.github.io/picard) and SAMtools [20] are required for deduplication and quality controls. The Genome Analysis Toolkit (GATK) [21] is adopted in our proposed pipeline by default to identify germline heterozygous loci. R packages SCALE, as well as its dependents—tsne (https://cran.r-project.org/package=tsne) and rje (https://cran.r-project.org/package=rje), is required for performing ASE analysis in R.

# 3 Methods

Figure 1 gives an overview of the analysis pipeline by SCALE. We start with bioinformatic preprocessing, which returns allele-specific read counts for the single cells. SCALE takes as input the allelic coverage at germline heterozygous loci and carries out three major steps: (1) gene classification using an empirical Bayes method; (2) estimation of allele-specific bursting kinetics using a Poisson–Beta hierarchical model with adjustment of technical variability; and (3) hypothesis testing of the two alleles of a gene to determine if they have different bursting kinetics and/or nonindependent bursting.

*3.1 Bioinformatics Pipeline*

For endogenous RNAs, germline heterozygous loci are called by bulk DNA- or RNA-seq of the same cells following the best practices for GATK [21]. If bulk sequencing from the same tissue is not available, scRNA-seq data can be aggregated to generate a pseudo-bulk RNA-seq sample. STAR [19] and BWA [18] are used for read alignment for RNA-seq and DNA-seq, respectively, followed by a deduplication and quality control procedure. We then force call single-cell allele-specific read counts using the mpileup command by SAMtools [20]. These allele-specific reads counts are further used as input to SCALE.

For external spike-ins, SCALE takes as input the true concentrations and lengths of the spike-in molecules, as well as the depth of coverage for each spike-in sequence across cells. The true concentration of each spike-in molecule is calculated according to the known concentration (denoted as $\mu$ attomoles/$\mu$L) and the dilution factor (e.g., $\times 40,000$):

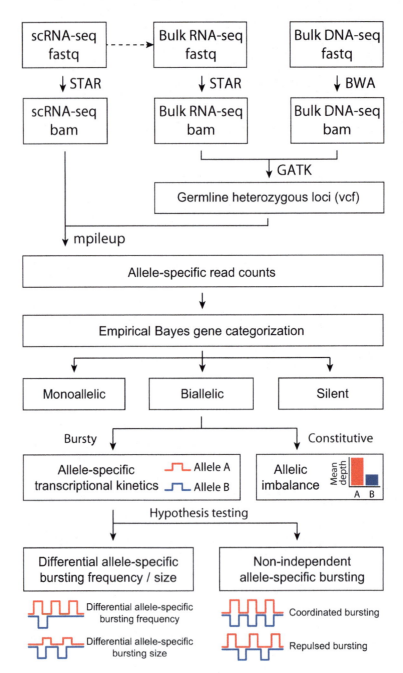

**Fig. 1** Overview of analysis pipeline of SCALE. SCALE takes as input allele-specific read counts at germline heterozygous loci and carries out three major steps: gene classification, estimation of allele-specific bursting kinetics, and hypothesis testing of differential and nonindependent allelic bursting

$$\frac{\mu \times 10^{-18}\, \text{moles}/\mu L \times 6.02214 \times 10^{23}\, \text{mole}^{-1}(\text{Avogadro  constant})}{40,000\, (\text{Dilution  factor})}.$$

The observed number of reads for each spike-in is calculated by adjusting for the library size factor, the read length, and the length of the spike-in RNA.

The bash scripts for bioinformatic preprocessing to generate the input for SCALE are available at https://github.com/yuchaojiang/SCALE/tree/master/bioinfo. Here, we outline the bioinformatic pipeline in three parts: (1) profile germline heterozygous loci by bulk DNA-seq; (2) profile allele-specific read counts at germline heterozygous loci by scRNA-seq; and (3) generate input for exogenous spike-ins for adjustment of technical variability.

```
################################################################
# 1. Profile germline heterozygous loci
################################################################
# 1.1. Index the genome template
bwa index ucsc.hg19.fasta
# 1.2. Align reads
bwa mem -M -t 16 ucsc.hg19.fasta bulk_R1.fastq bulk_R2.fastq >
bulk.sam
# 1.3. Convert sam to bam and sort
samtools view -bS bulk.sam > bulk.bam
java -jar SortSam.jar INPUT=bulk.bam OUTPUT=bulk.sorted.bam
SORT_ORDER=coordinate
# 1.4. Add read group
java -jar AddOrReplaceReadGroups.jar INPUT=bulk.sorted.bam
OUTPUT=bulk.sorted.rg.bam RGID=LANE1 RGLB=LIB1 RGPL=ILLUMINA
RGPU=UNIT1 RGSM=bulk
samtools index bulk.sorted.rg.bam
# 1.5. Dedup
java -jar MarkDuplicates.jar INPUT=bulk.sorted.rg.bam
OUTPUT=bulk.sorted.rg.dedup.bam
METRICS_FILE=bulk.sorted.rg.dedup.metrics.txt PROGRAM_RECORD_ID=
MarkDuplicates PROGRAM_GROUP_VERSION=null
PROGRAM_GROUP_NAME=MarkDuplicates
java -jar BuildBamIndex.jar INPUT= bulk.sorted.rg.dedup.bam
# 1.6. Profile germline heterozygous loci by GATK HaplotypeCaller
java -jar GenomeAnalysisTK.jar -R ucsc.hg19.fasta -T
HaplotypeCaller -I bulk.sorted.rg.dedup.bam -o
bulk.sorted.rg.dedup.raw.snps.indels.g.vcf
```

```
##################################################################
# 2. Profile allele-specific read counts by scRNA-seq
##################################################################
# 2.1. Get splice junction database:
wget http://labshare.cshl.edu/shares/gingeraslab/www-
data/dobin/STAR/STARgenomes/GENCODE/Old/gencode.v14.annotation.gt
f.sjdb
# 2.2. Generate the genome using STAR, 100bp paired-end
sequencing
genomeDir=directory_to_genome
STAR --runMode genomeGenerate --genomeDir $genomeDir --
genomeFastaFiles hg19.fa --sjdbFileChrStartEnd
gencode.v14.annotation.gtf.sjdb --sjdbOverhang 99 --runThreadN 4
# 2.3. Align reads
genomeDir=directory_to_genome
STAR --genomeDir genomeDir --readFilesIn samp_1.fastq
samp_2.fastq --outFilterIntronMotifs
RemoveNoncanonicalUnannotated --outFileNamePrefix samp_ --
runThreadN 4
# 2.4. Convert sam to bam, filter, and sort
samtools view -bS samp_Aligned.out.sam > samp_Aligned.out.bam
perl filter_sam_v2.pl samp_Aligned.out.bam
samp_Aligned.out.filtered.sam
samtools view -bS samp_Aligned.out.filtered.sam >
samp_Aligned.out.filtered.bam
java -Xmx30G -jar SortSam.jar INPUT=samp_Aligned.out.filtered.bam
OUTPUT=samp_Aligned.out.filtered.sorted.bam SORT_ORDER=coordinate
# 2.5. Add read group and index
java -Xmx30G -jar AddOrReplaceReadGroups.jar
```

```
INPUT=samp_Aligned.out.filtered.sorted.bam
OUTPUT=samp_Aligned.out.filtered.sorted.rg.bam RGID=LANE2
RGLB=LIB2 RGPL=ILLUMINA RGPU=UNIT2 RGSM=samp
samtools index samp_Aligned.out.filtered.sorted.rg.bam
# 2.6. Parse file: position.txt contains all the heterozygous
loci (chr + coordinate) returned by GATK HaplotypeCaller using
bulk DNA-seq
samtools mpileup -E -f hg19.fa -d 1000000 --position position.txt
samp_Aligned.out.filtered.sorted.rg.bam >
samp_Aligned.out.filtered.sorted.rg.mpileup
perl pileup2base_no_strand.pl
samp_Aligned.out.filtered.sorted.rg.mpileup 30
samp_Aligned.out.filtered.sorted.rg.parse30.txt
###################################################################
# 3. Generate input for exogenous spike-ins
###################################################################
# 3.1. Concatenate ERCC with genome (hg19) and index
cat ERCC92.fa hg19.fa > hg19_ERCC.fa
java -jar CreateSequenceDictionary.jar R= hg19_ERCC.fa O=
hg19_ERCC.dict
samtools faidx hg19_ERCC.fa
# 3.2. Generate the genome using STAR, 50bp paired-end sequencing
genomeDir=directory_to_ERCC_genome
STAR --runMode genomeGenerate --genomeDir $genomeDir --
genomeFastaFiles hg19_ERCC.fa --sjdbFileChrStartEnd
gencode.v14.annotation.gtf.sjdb --sjdbOverhang 99 --runThreadN 4
# 3.3. Align reads
genomeDir=directory_to_ERCC_genome
STAR --genomeDir genomeDir --readFilesIn samp_1.fastq
```

Meichen Dong and Yuchao Jiang

```
samp_2.fastq --outFilterIntronMotifs
RemoveNoncanonicalUnannotated --outFileNamePrefix samp_ --
runThreadN 4
# 3.4. Convert sam to bam, filter, and sort
samtools view -bS samp_Aligned.out.sam > samp_Aligned.out.bam
perl filter_sam_v2.pl samp_Aligned.out.bam
samp_Aligned.out.filtered.sam
samtools view -bS samp_Aligned.out.filtered.sam >
samp_Aligned.out.filtered.bam
java -Xmx30G -jar SortSam.jar INPUT=samp_Aligned.out.filtered.bam
OUTPUT=samp_Aligned.out.filtered.sorted.bam SORT_ORDER=coordinate
# 3.5. Add read group and index
java -Xmx30G -jar AddOrReplaceReadGroups.jar
INPUT=samp_Aligned.out.filtered.sorted.bam
OUTPUT=samp_Aligned.out.filtered.sorted.rg.bam RGID=LANE2
RGLB=LIB2 RGPL=ILLUMINA RGPU=UNIT2 RGSM=samp
samtools index samp_Aligned.out.filtered.sorted.rg.bam
# 3.6. Get total read counts as well as read counts for ERCC
samtools view -c samp_Aligned.out.filtered.sorted.rg.bam
while read ercc; do
echo $ercc
while read bam; do samtools view -c $bam $ercc; done <
rg.bam.list | cat > $ercc.txt
done < ercc.id
```

**3.2 Installation and Data Input**

The R package for SCALE can be installed directly from GitHub. The analysis by SCALE requires scRNA-seq data of cells from a homogenous cell population (i.e., from the same cell types and the same tissue). Each allele at heterozygous loci should have an expression matrix with rows as genes and columns as cells. In addition to the read count matrices for the endogenous RNAs, an input matrix for the spike-ins is needed for capturing technical variability. This matrix should have rows as spike-ins, the first column as the true number of molecules, the second column as the lengths of the molecules, and the third column and on as the observed read counts across cells. Note that spike-ins are not required for each individual cell (*see* **Note 1**). Here, we demonstrate the analysis framework of SCALE on a scRNA-seq data set of 122 mouse blastocyst cells from Deng et al. [2].

```
# SCALE installation
install.packages(c("rje", "tsne", "devtools"))
devtools::install_github("yuchaojiang/SCALE/package")
# Data input
library(SCALE)
data(mouse.blastocyst) # scRNA-seq dataset of mouse blastocyst
alleleA <- mouse.blastocyst$alleleA # Read counts for allele A
alleleB <- mouse.blastocyst$alleleB # Read counts for allele B
spikein_input <- mouse.blastocyst$spikein_input # Spike-in input
matrix
genename <- rownames(alleleA) # Rows correspond to genes
sampname <- colnames(alleleA) # Columns correspond to samples
# Input matrix for spike-ins
spikein_input[1:4,1:4]
             spikein_mol spikein_length GSM1112664 GSM1112665
RNA_SPIKE_1  12165.65657            755     462995     180148
1897542
RNA_SPIKE_2  12165.65657            755     336378     112809
1324062
RNA_SPIKE_3    912.42424           1003      18706       6062
88117
RNA_SPIKE_4    912.42424           1000      13689       6733
85575
```

**3.3  Quality Control**    Quality control (QC) procedures are recommended to filter out both poor-quality cells and extreme genes before applying SCALE. Cell QC metrics may include library size factor, which can be calculated by the following definition:

$$\eta_{c} = \underset{g}{\text{median}} \frac{Q_{cg}^{A} + Q_{cg}^{B}}{\left[ \Pi_{c*=1}^{C} \left( Q_{c*g}^{A} + Q_{c*g}^{B} \right) \right]^{1/C}},$$

where $\eta_{c}$ is the library size for cell c; $Q_{cg}^{A}$, $Q_{cg}^{B}$ denote the observed expression level of gene g from cell c for allele A and allele B, respectively; $C$ is the total number of cells. Cells with extreme ratios of reads that map to spike-ins versus endogenous genes should be excluded, which is equivalent to removing cells with extreme cell sizes.

SCALE needs to be applied to a homogeneous cell population, where the same bursting kinetics are shared across cells. Possible heterogeneity due to, for example, cell subgroups, developmental trajectories, and donor effects, can lead to biased downstream analysis. Therefore, it is strongly recommended to first perform dimensionality reduction and clustering method (e.g., PCA, t-SNE [22], ZIFA [23], SC3 [24], SIMLR [25], GiniClust2 [26], etc.) to the gene expression matrix. After clustering, remove cell outliers and apply SCALE to a homogeneous cell cluster with the assumption of shared bursting kinetics between the cells. (*See* **Note 2** for more details.) The input for SCALE is cell-type specific read count matrix for the two alleles, with rows being genes and columns being cells.

*3.4 Technical Variability*

To account for the tremendous amount technical variability observed in the scRNA-seq data, a hierarchical model based on the Toolkit for Analysis of Single Cell data (TASC) [15] is fit to the spike-in data. Specifically, let $Q_{cg}$ and $\Upsilon_{cg}$ be the observed and true expression levels of gene g in cell c, respectively. The hierarchical model used to model dropout, amplification, and sequencing bias is:

$$Q_{cg} \sim Z_{cg} \; \text{Poisson}\left(\alpha_c (\Upsilon_{cg})^{\beta_c}\right),$$

$$Z_{cg} \sim \text{Bernoulli}(\pi_{cg}),$$

$$\pi_{cg} = \text{expit}\left(\kappa_c + \tau_c \log(\Upsilon_{cg})\right),$$

where $Z_{cg}$ is a Bernoulli random variable indicating that gene g is detected in cell c. $\pi_{cg} = P(Z_{cg} = 1)$ is the success probability, which depends on $\log(\Upsilon_{cg})$, the logarithm of the true underlying expression. $\alpha_c$ models the capture and sequencing efficiency; $\beta_c$ models the amplification bias; $\kappa_c$ and $\tau_c$ characterize whether a transcript is successfully captured in the library.

$\{\alpha, \beta, \kappa, \tau\}$ are estimated through exogenous spike-ins and are assumed to be shared across cells from the same sequencing batch. $\alpha_c$ and $\beta_c$ are estimated by fitting a log-linear regression model. Nelder-Mead simplex algorithm is then applied to jointly optimize $\kappa_c$ and $\tau_c$, which models the probability of dropout events (*see* **Note 3**). The *tech_bias* function directly returns the estimated parameters and generates two plots by default, one for amplification bias and the other for dropout, as is shown in Fig. 2.

(A) Amplification and sequencing bias

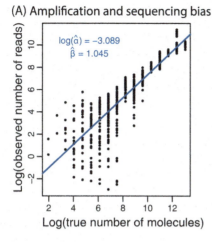

(B) **Dropout**

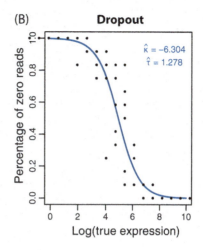

**Fig. 2** Modeling of technical variability and parameter estimation. Amplification and sequencing bias are modeled and captured by parameter $\alpha$ and $\beta$. Estimation is carried out by log-linear regression. Probability of dropout is modeled by $\kappa$ and $\tau$ and depends on the logarithm of the true expression. Estimation is carried out by the Nelder-Mead simplex algorithm

```
# Estimate parameters for technical variability
abkt <- tech_bias(spikein_input = spikein_input,
                  alleleA = alleleA, alleleB = alleleB,
                  readlength = 50, pdf = FALSE)
```

**3.5 Gene Classification**

SCALE adopts an empirical Bayes method that categorizes each gene into being silent, monoallelically expressed, and biallelically expressed based on their ASE across cells. An expectation maximization algorithm is implemented for fast estimation of the corresponding parameters. The result derived from the *gene_classify* function is a list of four elements: gene category, proportion of cells expressing allele A, proportion of cells expressing allele B, and posterior assignment of cells for each gene. For the posterior assignment for each gene in each cell, "A" corresponds to cells expressing A allele only, "B" corresponds to cells expressing B allele only, "AB" corresponds to cells expressing both alleles, and "Off" corresponds to cells that are silent for the gene of interest.

```
# Gene classification
gene.class.obj <- gene_classify(alleleA = as.matrix(alleleA),
                                alleleB = as.matrix(alleleB))
gene.category <- gene.class.obj$gene.category
```

**3.6 Allele-Specific Bursting Kinetics**

When studying ASE in single cells, it is critical to consider transcriptional bursting due to its pervasiveness in various organisms [27–30]. A two-state kinetic model has been proposed for gene transcription, where genes switch between ON and OFF states with activation rate $k_{on}$ and deactivation rate $k_{off}$. When a gene is at the ON state, DNA is transcribed to RNA at rate $s$, while RNA decays at rate $d$. A Poisson–Beta stochastic model for transcriptional bursting was proposed by Kepler and Elston [31]:

$$\Upsilon \sim \text{Poisson}(sp), \quad p \sim \text{Beta}(k_{on}, \ k_{off}),$$

where $\Upsilon$ is the number of RNA molecules and $p$ is the fraction of time that the gene spends in the active state. Note that the decay rate $d$ is set to 1 since only the stationary distribution is observed [32]. This Poisson–Beta model is easy to fit mathematically, with its parameters corresponding to biologically meaningful quantities—burst size as $s/k_{off}$ and burst frequency as $k_{on}$.

After gene classification, SCALE proceeds to infer allele-specific bursting parameters for biallelic bursty genes (*see* **Note 4**) using a hierarchical model:

$$\Upsilon_{cg}^{A} \sim \text{Poisson}\left(\phi_c s_g^A p_{cg}^A\right), \quad \Upsilon_{cg}^{B} \sim \text{Poisson}\left(\phi_c s_g^B p_{cg}^B\right),$$

$$p_{cg}^A \sim \text{Beta}\left(k_{on,g}^A, k_{off,g}^A\right), p_{cg}^B \sim \text{Beta}\left(k_{on,g}^B, k_{off,g}^B\right),$$

where $\Upsilon_{cg}^A$ and $\Upsilon_{cg}^B$ are the true ASE values for gene g in cell c. Note that the two Poisson–Beta distributions have gene- and allele-specific bursting parameters and share the same cell-size factor, which has been shown to affect burst size [33]. When spike-ins are available, cell size can be estimated by the ratio of the total number of endogenous RNA reads over the total number of spike-in reads [34]. Moreover, users can input the cell size factors $\phi_c$ if they are experimentally measured (*see* **Note 5** for details).

Since $\Upsilon_{cg}^A$ and $\Upsilon_{cg}^B$ are not directly observable while the observed ASE levels $Q_{cg}^A$ and $Q_{cg}^B$ are confounded with technical bias, we use a novel "histogram-repiling" method to derive the distribution of $\Upsilon_{cg}$ from the observed distribution $Q_{cg}$ for each gene. The allele-specific parameters $\left\{s^A, \ s^B, \ k_{on}^A, \ k_{on}^B, \ k_{off}^A, \ k_{off}^B\right\}$ are then estimated using the moment estimator methods.

For real dataset analysis, SCALE's function *allelic_kinetics* returns an object *allelic.kinetics.obj*, which contains the estimated bursting parameters. A pdf plot is generated by default and is shown in Fig. 3, where each dot corresponds to a gene and the genes off the diagonal indicate differential bursting kinetics between the two alleles.

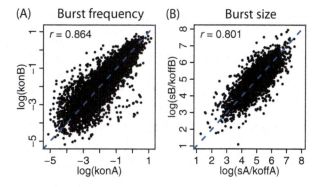

**Fig. 3** Allele-specific transcriptional kinetics of 7486 bursty genes from 122 blastocyst cells. (**a**) Burst frequency of the two alleles has a correlation of 0.864, where 485 genes show significant difference between the two alleles after FDR control. (**b**) Burst size of the two alleles has a correlation of 0.801, where 49 genes show significant difference between the two alleles

```
# Estimate bursting kinetics parameters
allelic.kinetics.obj <- allelic_kinetics(alleleA = alleleA, alleleB = alleleB,
                                abkt = abkt, gene.category = gene.category,
                                cellsize = cellsize, pdf = TRUE)
```

***3.7 Hypothesis Testing***

Nonparametric hypothesis testing is carried out with the null hypothesis that the two alleles share the same burst frequency and burst size $k_{on}^A = k_{on}^B$, $s^A/k_{off}^A = s^B/k_{off}^B$. SCALE's function *diff_allelic_bursting* has two "modes": the "raw" mode where the bootstrap samples from the raw observed read counts; and the "corrected" mode where the bootstrap samples from the adjusted allelic read counts. Two vectors of *p*-values will be obtained for test of burst frequency and burst size, respectively.

```
# Bootstrap based testing for bursting kinetics between two alleles
diff.allelic.obj <- diff_allelic_bursting(alleleA = alleleA, alleleB = alleleB,
                                cellsize = cellsize, gene.category =
                                gene.category, abkt = abkt,
                                allelic.kinetics.obj = allelic.kinetics.obj,
                                mode = 'corrected')
pval.kon <- diff.allelic.obj$pval.kon
pval.size <- diff.allelic.obj$pval.size
```

Chi-square test is carried out to test whether the two alleles burst independently. For genes where the proportion of cells expressing both alleles is significantly higher than expected, we define their bursting as coordinated; for genes where the proportion of cells expressing only one allele is significantly higher than expected, we define their bursting as repulsed. SCALE's function *non_ind_bursting* returns a list of two vectors, with one vector being the *p*-values and the other being the nonindependent bursting type ("C" as coordinated bursting and "R" as repulsed bursting).

```
# Chi-square test for fire dependence of the two alleles of a
gene
non.ind.obj <- non_ind_bursting(alleleA = alleleA,
alleleB = alleleB, gene.category = gene.category,
results.list = results.list)
```

**3.8  Plot and Output**     For each gene, a pdf plot can be generated with the estimated parameters and testing results. As an example, Fig. 4 shows the output by SCALE for gene *Hvcn1*, whose two alleles share similar burst size and frequency but burst in a coordinated fashion with nominal $p$-value less than 0.05.

```
# Generate a pdf output for a selected gene
i = which(genename == 'Hvcn1')
allelic_plot(alleleA = alleleA, alleleB = alleleB,
            gene.class.obj = gene.class.obj,
            allelic.kinetics.obj = allelic.kinetics.obj,
            diff.allelic.obj = diff.allelic.obj,
            non.ind.obj = non.ind.obj, i = i)
```

SCALE generates a final output as a tab delimited txt file. Table 1 includes output from a selected set of genes within each gene category as rows. The columns include: genename (gene name), gene. category (gene category), konA (burst frequency for allele A), konB (burst frequency for allele B), pval.kon ($p$-value from testing of shared burst frequency), sizeA (burst size for allele A), sizeB (burst size for allele B), pval.size ($p$-value from testing of shared burst size), A_cell, B_cell, AB_cell, Off_cell (number of cells with posterior assignment of A, B, AB, and Off), A_prop (proportion of cells expressing A allele), B_prop (proportion of cells expressing B allele), p.ind ($p$-value of burst independence), and non.ind.type (direction of non-independent bursting as coordinated or repulsed).

```
# Output all results from SCALE to a txt file
SCALE.output <- output_table(alleleA = alleleA, alleleB = alleleB,
                gene.class.obj = gene.class.obj,
                allelic.kinetics.obj = allelic.kinetics.obj,
                diff.allelic.obj = diff.allelic.obj,
                non.ind.obj = non.ind.obj)
```

At the whole genome level, we identified a significant number of genes whose allele-specific bursting differs according to burst

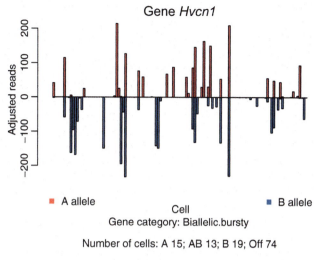

**Fig. 4** SCALE output for a bursty gene *Hvcn1*. Bar plot shows the adjusted allelic coverage across cells. Number of cells expressing A allele, B allele, both alleles, and neither allele is reported, together with the inferred allelic bursting kinetics. *P*-values from testing of shared bursting kinetics and nonindependent bursting are also returned

frequency but not burst size. Our findings provide evidence that allelic differences in the expression of bursty genes are achieved through differential modulation of burst frequency than burst size. Previous studies have shown that kinetic parameter that varies the most—along cell cycle [33], between different genes [35], between different growth conditions [36], or under regulation by a transcription factor [37]—is the probabilistic rate of switching to the active stat $k_{on}$, while the rates of gene inactivation $k_{off}$ and of transcription $s$ vary much less.

# 4   Notes

1. SCALE uses external spike-ins to estimate the parameters associated with technical noise but does not require spike-ins in every cell in the experiment. As long as *some* cells from the same batch have spike-ins, they can be used to model and capture the dropout events, as well as the amplification and sequencing bias during the library preparation and sequencing step. When spike-ins are not readily available, imputation based methods such as SAVER [16] and scImpute [17] can be adopted to recover the true underlying expression distribution.

**Table 1**
**Selected gene output from SCALE**

| Genename | Gene.category | konA | konB | pval.kon | sizeA | sizeB | pval.size | A_cell | B_cell | AB_cell | Off_cell | A.prop | B.prop | pval.ind | non.ind.type |
|---|---|---|---|---|---|---|---|---|---|---|---|---|---|---|---|
| *Phf7* | Biallelic.bursty | 0.0629 | 0.012 | 0.010 | 64.39 | 156.36 | 0.149 | 18 | 8 | 1 | 95 | 0.16 | 0.07 | 0.701 | R |
| *Zcchc4* | Biallelic.bursty | 0.0459 | 0.066 | 0.384 | 136.78 | 78.31 | 0.308 | 14 | 16 | 1 | 91 | 0.12 | 0.14 | 0.385 | R |
| *Slc16a5* | Biallelic.bursty | 0.0138 | 0.030 | 0.215 | 405.56 | 361.02 | 0.861 | 9 | 13 | 0 | 100 | 0.07 | 0.11 | 0.281 | R |
| *Gprc5a* | Biallelic.bursty | 0.5515 | 0.810 | 0.131 | 578.29 | 644.77 | 0.783 | 12 | 21 | 80 | 5 | 0.78 | 0.86 | 0.428 | C |
| *Atp6ap2* | Biallelic.bursty | 1.0221 | 0.031 | 0.000 | 141.82 | 1135.68 | 0.000 | 90 | 2 | 13 | 17 | 0.84 | 0.12 | 0.798 | C |
| *Nfx1* | Biallelic.bursty | 0.3342 | 0.377 | 0.672 | 121.29 | 233.68 | 0.236 | 13 | 31 | 48 | 24 | 0.53 | 0.68 | 0.010 | C |
| *Rpl4* | Biallelic.nonbursty | – | – | – | – | – | – | 0 | 0 | 122 | 0 | 1.00 | 1.00 | – | – |
| *Exosc8* | Biallelic.nonbursty | – | – | – | – | – | – | 2 | 4 | 116 | 0 | 0.97 | 0.98 | – | – |
| *Arhgap1* | Biallelic.nonbursty | – | – | – | – | – | – | 9 | 4 | 108 | 0 | 0.97 | 0.93 | – | – |
| *Gla* | MonoA | – | – | – | – | – | – | 13 | 0 | 0 | 109 | 0.11 | 0.00 | – | – |
| *Agbl2* | MonoA | – | – | – | – | – | – | 2 | 0 | 0 | 120 | 0.02 | 0.00 | – | – |
| *Espn* | MonoA | – | – | – | – | – | – | 1 | 0 | 0 | 121 | 0.01 | 0.00 | – | – |
| *Rasd2* | MonoB | – | – | – | – | – | – | 0 | 7 | 0 | 115 | 0.00 | 0.06 | – | – |
| *Spata2L* | MonoB | – | – | – | – | – | – | 0 | 4 | 0 | 118 | 0.00 | 0.03 | – | – |
| *Kcnip2* | MonoB | – | – | – | – | – | – | 0 | 2 | 0 | 120 | 0.00 | 0.02 | – | – |
| *Gm101* | Silent | – | – | – | – | – | – | 0 | 0 | 0 | 122 | 0.00 | 0.00 | – | – |
| *Polr2l* | Silent | – | – | – | – | – | – | 0 | 0 | 0 | 122 | 0.00 | 0.00 | – | – |
| *Actb* | Silent | – | – | – | – | – | – | 0 | 0 | 0 | 122 | 0.00 | 0.00 | – | – |

The final output of SCALE is a tab delimited text file. The columns include: gene name, gene category, bursty frequency A, burst frequency B, $p$-value of shared burst frequency, burst size A, burst size B, $p$-value of shared burst size, number of cells with posterior assignment of A, B, AB, and null, proportion of cells expressing A allele, proportion of cells expressing B allele, $p$-value of burst independence, and direction of non-independent bursting (C for coordinated bursting and R for repulsed bursting)

2. SCALE needs to be applied to a *homogeneous* cell population, where the same bursting kinetics are shared across all cells. Possible heterogeneity due to, for example, cell subgroups, lineages, and donor effects, can lead to biased downstream analysis. We find that an excessive number of significant genes showing coordinated bursting between the two alleles can be indicative of heterogeneity with the cell population, which should be further stratified. Therefore, it is recommended that users adopt dimensionality reduction and clustering methods (e.g., PCA, t-SNE [22], ZIFA [23], SC3 [24], SIMLR [25], GiniClust2 [26], etc.) before applying SCALE. The input data for SCALE should be cluster-specific allelic read counts across all genes in all cells.

3. Sometimes there can be a "poor" fitting of $\kappa$ and $\tau$ due to low sequencing depth (small $\alpha$) as well as low amplification efficiency (small $\beta$) in the library preparation and sequencing procedure. This results in $\alpha \times (\Upsilon^{\beta})$ being much smaller than $\Upsilon$ and sometimes significantly smaller than 1 for spike-ins with low concentrations and endogenous genes with low expressions. In this case, whether there is dropout or not (modeled by $\kappa$ and $\tau$), the observed expression will be zero from Poisson sampling when sequencing is carried out. As such, $\kappa$ and $\tau$ are not statistically identifiable. Empirical evidence showed that this issue does not affect the downstream analysis of bursting kinetics, since the lowly expressed transcripts resulting in zero read counts will be modeled and captured, whether it is due to Poisson sampling or dropout.

4. Through the empirical Bayes framework, a gene is categorized by SCALE to be monoallelically expressed, if only one allele is expressed in a nonzero proportion of cells. While SCALE is focused on detecting differential bursting kinetics between the two alleles, monoallelic expression is an extreme case where one allele is completely off—This is equivalent to the allele having an infinitely large burst frequency. We do not infer bursting kinetics on these monoallelically expressed genes since there is no way nor need to do so. Nevertheless, the gene categorization results return genes with monoallelic expression (*see* Table 1 for an example) and can be used for downstream analysis.

5. It is nontrivial to reliably estimate the cell size factor in silico. Cell size can be inferred from the expression levels of housekeeping genes such as *Gapdh* [33] or from the ratio of the total number of reads mapped to the endogenous RNA and the total number of reads mapped the spike-ins [34]. These two measurements are not on the same scale and should not be combined. In real dataset analysis when experimentally measured cell sizes are not available, we recommend first estimating the bursting kinetics with all cell size factors set to one.

After this, one can try again using the in silico estimated cell size factors. The correlation between the bursting kinetics of the two alleles can serve as a sanity check for good data quality and accurate cell size estimation—On the genome-wide scale, they should be correlated with a decent correlation coefficient.

## Acknowledgment

This work was supported by NIH grant CA142538 and a developmental award from the UNC Lineberger Comprehensive Cancer Center 2017T109 (to YJ). We thank Dr. Nancy R Zhang and Dr. Mingyao Li for helpful comments and suggestions.

## References

1. Buckland PR (2004) Allele-specific gene expression differences in humans. Hum Mol Genet 13(2):R255–R260. https://doi.org/10.1093/hmg/ddh227

2. Deng Q, Ramskold D, Reinius B, Sandberg R (2014) Single-cell RNA-seq reveals dynamic, random monoallelic gene expression in mammalian cells. Science 343:193–196. https://doi.org/10.1126/science.1245316

3. Reinius B, Sandberg R (2015) Random monoallelic expression of autosomal genes: stochastic transcription and allele-level regulation. Nat Rev Genet 16:653–664. https://doi.org/10.1038/nrg3888

4. Reinius B, Mold JE, Ramskold D, Deng Q, Johnsson P, Michaelsson J, Frisen J, Sandberg R (2016) Analysis of allelic expression patterns in clonal somatic cells by single-cell RNA-seq. Nat Genet 48:1430–1435. https://doi.org/10.1038/ng.3678

5. Skelly DA, Johansson M, Madeoy J, Wakefield J, Akey JM (2011) A powerful and flexible statistical framework for testing hypotheses of allele-specific gene expression from RNA-seq data. Genome Res 21:1728–1737. https://doi.org/10.1101/gr.119784.110

6. Leon-Novelo LG, McIntyre LM, Fear JM, Graze RM (2014) A flexible Bayesian method for detecting allelic imbalance in RNA-seq data. BMC Genomics 15:920. https://doi.org/10.1186/1471-2164-15-920

7. Jiang Y, Zhang NR, Li M (2017) SCALE: modeling allele-specific gene expression by single-cell RNA sequencing. Genome Biol 18(1):74. https://doi.org/10.1186/s13059-017-1200-8

8. Benitez JA, Cheng S, Deng Q (2017) Revealing allele-specific gene expression by single-cell transcriptomics. Int J Biochem Cell Biol 90:155–160. https://doi.org/10.1016/j.biocel.2017.05.029

9. Kim JK, Marioni JC (2013) Inferring the kinetics of stochastic gene expression from single-cell RNA-sequencing data. Genome Biol 14:R7. https://doi.org/10.1186/gb-2013-14-1-r7

10. Levesque MJ, Ginart P, Wei Y, Raj A (2013) Visualizing SNVs to quantify allele-specific expression in single cells. Nat Methods 10:865–867. https://doi.org/10.1038/nmeth.2589

11. Goetz JJ, Trimarchi JM (2012) Transcriptome sequencing of single cells with smart-Seq. Nat Biotechnol 30(8):763–765. https://doi.org/10.1038/nbt.2325

12. Picelli S, Faridani OR, Bjorklund AK, Winberg G, Sagasser S, Sandberg R (2014) Full-length RNA-seq from single cells using smart-seq2. Nat Protoc 9(1):171–181. https://doi.org/10.1038/nprot.2014.006

13. Macosko EZ, Basu A, Satija R, Nemesh J, Shekhar K, Goldman M, Tirosh I, Bialas AR, Kamitaki N, Martersteck EM, Trombetta JJ, Weitz DA, Sanes JR, Shalek AK, Regev A, McCarroll SA (2015) Highly parallel genome-wide expression profiling of individual cells using Nanoliter droplets. Cell 161(5):1202–1214. https://doi.org/10.1016/j.cell.2015.05.002

14. Zheng GX, Terry JM, Belgrader P, Ryvkin P, Bent ZW, Wilson R, Ziraldo SB, Wheeler TD, McDermott GP, Zhu J, Gregory MT, Shuga J, Montesclaros L, Underwood JG, Masquelier DA, Nishimura SY, Schnall-Levin M, Wyatt PW, Hindson CM, Bharadwaj R, Wong A, Ness KD, Beppu LW, Deeg HJ, McFarland C, Loeb KR, Valente WJ, Ericson NG, Stevens

EA, Radich JP, Mikkelsen TS, Hindson BJ, Bielas JH (2017) Massively parallel digital transcriptional profiling of single cells. Nat Commun 8:14049. https://doi.org/10.1038/ncomms14049

15. Jia C, Hu Y, Kelly D, Kim J, Li M, Zhang NR (2017) Accounting for technical noise in differential expression analysis of single-cell RNA sequencing data. Nucleic Acids Res 45 (19):10978–10988. https://doi.org/10.1093/nar/gkx754

16. Huang M, Wang J, Torre E, Dueck H, Shaffer S, Bonasio R, Murray JI, Raj A, Li M, Zhang NR (2018) SAVER: gene expression recovery for single-cell RNA sequencing. Nat Methods 15(7):539–542. https://doi.org/10.1038/s41592-018-0033-z

17. Li WV, Li JJ (2018) An accurate and robust imputation method scImpute for single-cell RNA-seq data. Nat Commun 9(1):997. https://doi.org/10.1038/s41467-018-03405-7

18. Li H, Durbin R (2009) Fast and accurate short read alignment with burrows-wheeler transform. Bioinformatics 25(14):1754–1760. https://doi.org/10.1093/bioinformatics/btp324

19. Dobin A, Davis CA, Schlesinger F, Drenkow J, Zaleski C, Jha S, Batut P, Chaisson M, Gingeras TR (2013) STAR: ultrafast universal RNA-seq aligner. Bioinformatics 29(1):15–21. https://doi.org/10.1093/bioinformatics/bts635

20. Li H, Handsaker B, Wysoker A, Fennell T, Ruan J, Homer N, Marth G, Abecasis G, Durbin R, Genome Project Data Processing Subgroup (2009) The sequence alignment/map format and SAMtools. Bioinformatics 25 (16):2078–2079. https://doi.org/10.1093/bioinformatics/btp352

21. DePristo MA, Banks E, Poplin R, Garimella KV, Maguire JR, Hartl C, Philippakis AA, del Angel G, Rivas MA, Hanna M, McKenna A, Fennell TJ, Kernytsky AM, Sivachenko AY, Cibulskis K, Gabriel SB, Altshuler D, Daly MJ (2011) A framework for variation discovery and genotyping using next-generation DNA sequencing data. Nat Genet 43(5):491–498. https://doi.org/10.1038/ng.806

22. van der Maaten L, Hinton G (2008) Visualizing data using t-SNE. J Mach Learn Res 9:2579–2605

23. Pierson E, Yau C (2015) ZIFA: dimensionality reduction for zero-inflated single-cell gene expression analysis. Genome Biol 16:241. https://doi.org/10.1186/s13059-015-0805-z

24. Kiselev VY, Kirschner K, Schaub MT, Andrews T, Yiu A, Chandra T, Natarajan KN, Reik W, Barahona M, Green AR, Hemberg M (2017) SC3: consensus clustering of single-cell RNA-seq data. Nat Methods 14(5):483–486. https://doi.org/10.1038/nmeth.4236

25. Wang B, Zhu J, Pierson E, Ramazzotti D, Batzoglou S (2017) Visualization and analysis of single-cell RNA-seq data by kernel-based similarity learning. Nat Methods 14(4):414–416. https://doi.org/10.1038/nmeth.4207

26. Tsoucas D, Yuan GC (2018) GiniClust2: a cluster-aware, weighted ensemble clustering method for cell-type detection. Genome Biol 19(1):58. https://doi.org/10.1186/s13059-018-1431-3

27. Chong S, Chen C, Ge H, Xie XS (2014) Mechanism of transcriptional bursting in bacteria. Cell 158:314–326. https://doi.org/10.1016/j.cell.2014.05.038

28. Blake WJ, Balazsi G, Kohanski MA, Isaacs FJ, Murphy KF, Kuang Y, Cantor CR, Walt DR, Collins JJ (2006) Phenotypic consequences of promoter-mediated transcriptional noise. Mol Cell 24:853–865. https://doi.org/10.1016/j.molcel.2006.11.003

29. Fukaya T, Lim B, Levine M (2016) Enhancer control of transcriptional bursting. Cell 166:358–368. https://doi.org/10.1016/j.cell.2016.05.025

30. Suter DM, Molina N, Gatfield D, Schneider K, Schibler U, Naef F (2011) Mammalian genes are transcribed with widely different bursting kinetics. Science 332:472–474. https://doi.org/10.1126/science.1198817

31. Kepler TB, Elston TC (2001) Stochasticity in transcriptional regulation: origins, consequences, and mathematical representations. Biophys J 81:3116–3136. https://doi.org/10.1016/s0006-3495(01)75949-8

32. Stegle O, Teichmann SA, Marioni JC (2015) Computational and analytical challenges in single-cell transcriptomics. Nat Rev Genet 16:133–145. https://doi.org/10.1038/nrg3833

33. Padovan-Merhar O, Nair GP, Biaesch AG, Mayer A, Scarfone S, Foley SW, Wu AR, Churchman LS, Singh A, Raj A (2015) Single mammalian cells compensate for differences in cellular volume and DNA copy number through independent global transcriptional mechanisms. Mol Cell 58:339–352. https://doi.org/10.1016/j.molcel.2015.03.005

34. Vallejos CA, Marioni JC, Richardson S (2015) BASiCS: Bayesian analysis of single-cell sequencing data. PLoS Comput Biol 11: e1004333. https://doi.org/10.1371/journal.pcbi.1004333

35. Skinner SO, Xu H, Nagarkar-Jaiswal S, Freire PR, Zwaka TP, Golding I (2016) Single-cell analysis of transcription kinetics across the cell cycle. Elife 5:e12175. https://doi.org/10.7554/eLife.12175

36. Ochiai H, Sugawara T, Sakuma T, Yamamoto T (2014) Stochastic promoter activation affects Nanog expression variability in mouse embryonic stem cells. Sci Rep 4:7125. https://doi.org/10.1038/srep07125

37. Xu H, Sepulveda LA, Figard L, Sokac AM, Golding I (2015) Combining protein and mRNA quantification to decipher transcriptional regulation. Nat Methods 12:739–742. https://doi.org/10.1038/nmeth.3446

# Chapter 12

## Using BRIE to Detect and Analyze Splicing Isoforms in scRNA-Seq Data

### Yuanhua Huang and Guido Sanguinetti

**Abstract**

Single-cell RNA-seq (scRNA-seq) provides a comprehensive measurement of stochasticity in transcription, but the limitations of the technology have prevented its application to dissect variability in RNA processing events such as splicing. In this chapter, we review the challenges in splicing isoform quantification in scRNA-seq data and discuss BRIE (Bayesian regression for isoform estimation), a recently proposed Bayesian hierarchical model which resolves these problems by learning an informative prior distribution from sequence features. We illustrate the usage of BRIE with a case study on 130 mouse cells during gastrulation.

**Key words** Alternative splicing, Isoform quantification, Single-cell RNA-seq, Bayesian model

## 1 Introduction

Next generation sequencing (NGS) technologies have had a huge impact on our understanding of biology, shedding unprecedented light on the role of genomic, epigenomic, and transcriptomic processes within cellular function. Recently, efficient RNA amplification techniques have been coupled with NGS to yield transcriptome sequencing protocols that can measure the abundance of transcripts within single cells, known as single-cell RNA-seq (scRNA-seq) [1]. scRNA-seq has provided unprecedented opportunities to investigate the stochasticity of transcription and its importance in cellular diversity. Groundbreaking applications of scRNA-seq include the ability to discover novel cell types, e.g., rare intestinal cell types [2] and distinct immune cell subtypes in health and disease conditions [3, 4], to reconstruct the trajectory of embryonic cells development [5, 6] or the fate of immune cells [7, 8], and to dissect the heterogeneity of tumor cells [9] or the complexity of its ecosystem [10]. This promising technology is being applied in many more studies in a wide range of cellular biology problems.

Guo-Cheng Yuan (ed.), *Computational Methods for Single-Cell Data Analysis*, Methods in Molecular Biology, vol. 1935, https://doi.org/10.1007/978-1-4939-9057-3_12, © Springer Science+Business Media, LLC, part of Springer Nature 2019

The vast majority of studies employing scRNA-seq probe individual cells at the *gene level*, aggregating all reads mapping to an annotated genic region into a quantification of expression for that gene. However, the biologically relevant output of transcription is the transcript, not the gene; multiple levels of regulation and processing, including RNA splicing, capping, and polyadenylation, are inevitably missed by a gene-level quantification. In this chapter we focus on splicing, the processing step where intronic regions are erased and exonic regions are joined to form the mature mRNA. Splicing is a key regulatory step in gene expression in most eukaryotes. In human, over 90% of genes have multiple splicing isoforms [11], often being required to express the right splicing isoform in a specific condition. Many studies have shown that alternative splicing plays an important role in cell differentiation, tissue identity, and organ development [12], for example over 300 genes express differential splicing isoforms between mouse embryonic and adult brain, including some main regulatory genes in nervous system [13]. Also, mis-splicing often leads to serious diseases, which may be caused by genetic mutations of splicing regulatory sequences, or the dysregulation of the spliceosome or specific splicing factors [14].

Despite its importance, splicing quantification has seldom been attempted in scRNA-seq studies. Technical challenges, including the minute amounts of starting material, low cDNA conversion efficiency, and uneven transcript coverage, mean that scRNA-seq data sets can exhibit substantial technical variation [15]. Additionally, the high level of dropout, and the generally lower coverage, pose significant problems to standard splicing quantification methods from bulk RNA-seq.

Nevertheless, the last few years saw a few studies attempting to stretch single-cell sequencing technologies to gain insights into RNA splicing. Faigenbloom et al. [16] studied the variability of the cassette exon inclusion across single cells, and highlighted the potential regulation of the evolutionary conservation of flanking introns and the correlation with its expression level. This study, however, did not attempt quantification of alternative splicing proportions within individual cells, merely highlighting the evidence from the data for different isoform production between different cells. Song and colleagues [17] used an exceptionally high (for single cells) sequencing depth to show that splicing can be an independent feature of cell identity in neuron cell differentiation, and also provided a software suite for analyzing splicing in single cells. More recent papers proposed the use of RNA splicing dynamics to predict the future state of individual cells [ ], and to investigate the role of epigenetic variability in splicing control [19].

In this Chapter, we describe BRIE, the first and one of the very few methods for splicing quantification from scRNA-seq [20]. BRIE is a Bayesian method which integrates sequence features with scRNA-seq data to obtain more confident quantifications of splicing ratios at low coverage. The method can be used both for splicing quantification and to detect differential splicing between different cells/ groups of cells. Additionally, the integrative aspects of BRIE can be used to quantify the effects of covariates (such as sequence or epigenetic status) on splicing, as done in [19]. BRIE is implemented as a stand-alone python package which is freely available at https://pypi.org/project/brie.

The rest of this chapter is organized as follows. In the Materials section, we briefly describe the software package and the preprocessed splicing annotation data. Then in the Methods section, we provide a brief and self-contained introduction to the methodological foundations of BRIE, followed by a tutorial introduction to its usage on a case study on 130 mouse cells during gastrulation [5]. A few notes (*see* **Notes 1–4**) on using BRIE are also provided.

## 2  Materials

The main software we need for this Chapter is BRIE, which is implemented as open source software in Python, compatible with both Python2 and Python3. BRIE was mainly designed for exon-skipping events: we provide preprocessed splicing events annotations for human and mouse and their according sequence features. In addition, BRIE has a separate toolkit package, BRIE-kit, to annotate and filter exon-skipping events and extract sequence features for any species. Both the software and preprocessed annotation data sets are freely available at following links,

1. BRIE package: https://pypi.org/project/brie.

2. BRIE-kit for preprocessing: https://pypi.org/project/briekit.

3. Preprocessed annotation datasets: https://sourceforge.net/ projects/brie-rna/files/annotation/.

In the next section, we will go through a case study on 130 mouse cells during gastrulation, which is a published scRNA-seq dataset produced using the SMART-seq2 protocol [5]. The original dataset consists of 1205 cells; in this tutorial we use a subset of cells for convenience: all the analyses described in this chapter can be carried out on a single small server within a few hours. For bigger dataset with thousands of cells, cluster with multiple nodes can be very useful to reduce the computational time. As the input files are specified in each subsection, this example pipeline is also applicable for any other customized dataset, and all command lines in bash files are included in BRIE's GitHub repository: https:// github.com/huangyh09/brie/tree/master/example/ gastrulation.

## 3    Methods

### 3.1   High-Level Model Description

Low coverage is a common problem for scRNA-seq data, and particularly brings a big statistical challenge to splicing quantification, as short sequenced RNA-seq reads can be aligned to multiple isoforms, namely not immediately informative on splicing status, and therefore it normally requires a high coverage for accurate estimate. Recent work has shown that improved predictions at lower coverage can be achieved with Bayesian methods by incorporating informative prior distributions within the probabilistic splicing quantification algorithms, leveraging either aspects of the experimental design, such as time series [21], or auxiliary data sets such as measurements of PolII localization [22]. In addition, it has also been demonstrated that splicing (in bulk cells) can be accurately predicted from sequence-derived features [23]. This suggests that overall patterns of read distribution may be associated with specific sequence words, so that one may be able to construct sequence-based informative prior distributions that may be learned directly from data. This is the idea at the core of BRIE (Bayesian Regression for Isoform Estimation), a statistical model that achieves extremely high sensitivity at low coverage by the use of informative priors learned directly from data via a (latent) regression model. The regression model couples the task of splicing quantification across different genes, allowing a statistical transfer of information from well-covered genes to lower covered genes, achieving considerable robustness to noise in low coverage.

Figure 1 presents a schematic illustration of BRIE (see Methods in the original paper for precise definitions and details of the estimation procedure). The bottom part of the figure represents the standard mixture model approach to isoform estimation introduced in MISO [24] and Cufflinks [25] and also used in many recent methods, e.g., Kallisto [26], where reads are associated with a latent, multinomially distributed isoform identity variable. This module takes as input the scRNA-seq data (aligned reads) and forms the likelihood of our Bayesian model. The multinomial identity variables are then assigned an informative prior in the form of a regression model (top half of Fig. 1), where the prior probability of inclusion ratios is regressed against sequence-derived features. Crucially, the regression parameters are shared across all genes and can be learned across multiple single cells, thus regularizing the task and enabling robust predictions in the face of very low coverage. While the class of regression models we employ is different from the neural networks of [23], they still provide a highly accurate supervised learning predictor of splicing on bulk RNA-seq data sets.

This architecture effectively enables BRIE to simultaneously trade-off two tasks: in the absence of data (drop-out genes), the

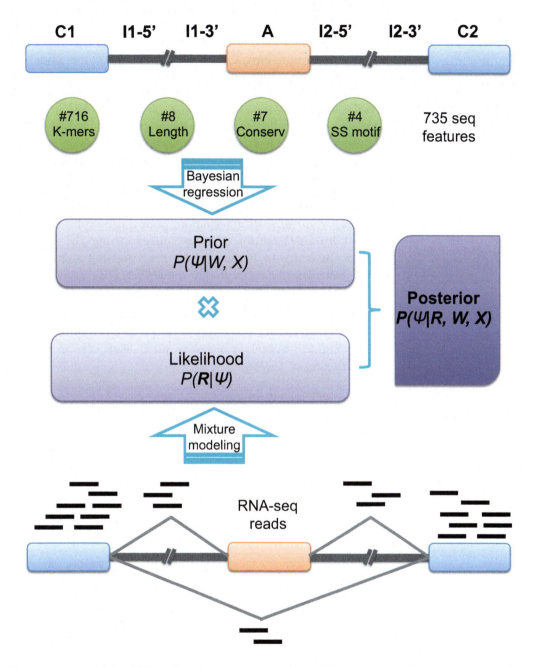

**Fig. 1** A cartoon of the BRIE method for isoform estimation. BRIE combines a likelihood computed from RNA-seq data (bottom part) and an informative prior distribution learned from 735 sequence-derived features (top)

informative prior provides a way of imputing missing data, while for highly covered genes the likelihood term dominates, returning a mixture-model quantification. For intermediate levels of coverage, BRIE uses Bayes's theorem to trade off imputation and quantification. A full description of the statistical model, as well as empirical results on both real and simulated data sets, is available in the original paper [20]. In the rest of this section, we focus on giving a thorough introduction to the usage of the associated software package.

### 3.2 Data Preparation with BRIE-Kit

In order to quantify the exon-skipping splicing with BRIE, we need a set of good quality annotated splicing events, and also their according sequence features. Besides using our preprocessed annotations for human and mouse, BRIE-kit, a separate Python package developed under the Python2 environment, can prepare the annotation for more species, through three functions (1) *briekit-event* for extracting the exon-skipping events from full gene annotation, (2) *briekit-event-filter* for filtering out poor quality splicing events, and (3) *briekit-factor* for defining and extracting the sequence features. In order to perform these preprocessing steps on the mouse gastrulation data (stored in the DATA_DIR directory), the following command lines would be used

1. Generate exon-skipping events from gene annotation.

```
$ briekit-event -a gencode.vM17.annotation.gtf.gz -o
$DATA_DIR/AS_events
```

2. Filter splicing events.

```
$ briekit-event-filter -a AS_events/SE.gff3.gz --anno_ref
gencode.vM17.annotation.gtf.gz \
-r GRCm38.p6.genome.fa
```

3. Extract the sequence features.

```
$ briekit-factor -a AS_events/SE.filtered.gff3.gz -r
GRCm38.p6.genome.fa \
-c mm10.60way.phastCons.bw -o mouse_features.csv -p 10 --
bigWigSummary ./bigWigSummary
```

This retrieves the 12,115 initial exon-skipping events; 5763 out of them pass the filtering conditions listed in the original BRIE paper [20]. The filtered annotation will be located in SE.filtered.gff3.gz, as a default output file. Also, the same set of 735 sequence features that are used in the original paper [20] for learning informative priors will be generated into the output file mouse_features.csv.gz. These two files will be used for splicing quantification in BRIE.

More detailed instructions, including downloading genome references and dependent software, can be found in the BRIE-kit GitHub wiki page (https://github.com/huangyh09/briekit/wiki)

and bash file for this example (https://github.com/huangyh09/briekit/blob/master/example/anno_mouse.sh).

### 3.3 Splicing Quantification Using BRIE

Once we obtained a set of exon-skipping events and their according sequence features, we can use BRIE to quantify their inclusion probabilities from scRNA-seq data. First, we need to download the raw scRNA-seq reads in fastq format, and align the reads to the genome, by HISAT [27] or STAR [28] or other splice-aware aligners. Then, for each cell, there will be a sorted and indexed alignment file in bam/sam format, e.g., cell_n.sorted.bam.

Now, we can run BRIE to quantify the annotated exon-skipping events on the aligned reads files by following command line,

```
$ brie -a SE.filtered.gff3.gz -s cell_n.sorted.bam -f mouse_-
features.csv.gz -o $OUT_DIR -p 10
```

BRIE then generates three output files in the OUT_DIR: fractions.tsv for exon inclusion fractions, samples.csv.gz for all MCMC samples, and weights.tsv for learned weights of these sequence predictors.

All the above procedures work on an individual cell. With these settings, it takes around 4 min for BRIE to quantify one cell on a single server. Similarly, given 10 nodes in parallel, 1300 cells can be done around 8 h. Naturally, the running time will depend on the specifications of the server, as well as on the number of splicing events quantified.

In addition, BRIE provides a simple way to pool all reads from a group of cells as an ensemble sample, which can be useful either for estimating a single-homogeneous splicing value in a group of cells or for learning common regression weights for the input features. The homogeneous splicing value in a group can be used for detecting differential splicing between groups, and the regression weights can be used as shared weights to (re-)quantify at single-cell resolution. For inputting multiple cells, the above command line just needs to list all cells with comma separation, and for setting shared weights for individual cells it is just by using -w, as follows,

```
$ brie -a SE.filtered.gtf -f mouse_features.csv.gz \
-s cell1.sorted.bam,cell2.sorted.bam,cell3.sorted.bam -o
$OUT_DIR/cell_1to4 -p 10
$ brie -a SE.filtered.gtf -f mouse_features.csv.gz \
-s cell1.sorted.bam -w $OUT_DIR/cell_1to4/weights.tsv -o
$OUT_DIR/cell1_fixedW -p 10
```

### 3.4 Differential Splicing Analysis with BRIE

Besides splicing quantification, a major focus of many studies is to detect the differential splicing between individual cells or between cell clusters. BRIE uses a Bayes factor to detect differential splicing between any pair of cells (or cell clusters), and the command line to run it as follows,

```
# a few cells in command line
$ fileList=cell1/samples.csv.gz,cell2/samples.csv.gz,cell3/
samples.csv.gz,cell4/samples.csv.gz
$ brie-diff -i $fileList -o cell1_cell4.diff.tsv
```

Then *brie-diff* will output two tsv files. The first file, in the format of xxx.diff.tsv, contains all pairs of cells (or cell groups) that passed the Bayes factor threshold. The other one, in the format of xxx.diff.rank.tsv, ranks the splicing events by the number of differentially spliced cell pairs, which can be used to select splicing marker for cell type identity. When using *brie-diff* to detect differential splicing events between two groups of cells, we provide some discussion in **Note 1**.

***3.5  Plotting Results and Extracting Statistics from the BRIE Output***

Once a set of highly variable splicing events has been detected, it is very useful to visualize the raw reads and the quantification results. Sashimi plots [29], which were originally developed by Yarden Katz and colleagues, are a visually effective way to display reads densities and junction reads. In BRIE, we adapted the sashimi plot to visualize the results, including the reads density and the prior and posterior distribution of splicing fraction. Sashimi_plot is included in BRIE-kit GitHub repository (https://github.com/huangyh09/briekit/tree/master/sashimi_plot), as a self-contained folder, which can be executed as follows,

```
SASHIMI=~/MyGit/briekit/sashimi_plot/sashimi_plot.py
python $SASHIMI --plot-event ENSMUSG00000027478.AS2 $GFF_DIR
sashimi_setting.txt \
--output-dir $PLOT_DIR --plot-label DNMT3B-exon2.pdf --plot-
title DNMT3B-exon2
```

Figure 2 shows an example of the type of plot that will be produced. Note, sashimi_plot is not part of the standard BRIE-kit package, but the above folder includes all scripts and demos to generate the sashimi plot. A few dependent Python packages are needed, e.g., misopy and matplotlib, and it is only compatible with Python2 but not Python3.

# 4  Notes

BRIE provides an effective solution to the problem of splicing quantification in single cells, by learning a sequence-based informative prior that can robustify results against low coverage. In this Chapter, we provided a tutorial introduction to the usage of the

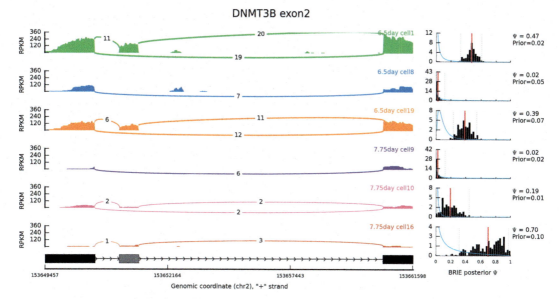

**Fig. 2** Visualization of splicing quantification with sashimi plot and histogram. An example exon-skipping event in DNMT3B in 3 mouse cells at 6.5 s days and 3 cells at 7.75 s days. The left panel is sashimi plot of the reads density and the number of junction reads. The right panel is the prior distribution in blue curve and a histogram of the posterior distribution in black, both learned by BRIE. For the histogram, the red line is the mean and the dash lines are the 95% confidence interval

BRIE python package; this is aimed at the practitioner bioinformatician that might want to replicate splicing analyses on scRNA-seq data. While we believe BRIE is a worthwhile addition to the armory of scRNA-seq statistical methods, some considerations as to its usage and possible extension are in order.

1. For detecting differential splicing between two cell groups in Subheading 3.4, there are two options. The first option is pooling reads from all cells within a group, and quantify splicing as one ensemble sample, which can be simply done by listing all cells with comma separation in *brie*. Then *brie-diff* can detect the differential splicing events between any pair of ensemble samples. This option assumes a homogeneous splicing level across the group, and it requires an additional splicing quantification besides the routine cell-level quantifications. Alternatively, one may use BRIE to quantify splicing in individual cells in each group, and then use a standard hypothesis test, e.g., Wilcoxon, to detect significant changes in the point estimates between two cell populations. This procedure requires however a sufficient number of cells in each population to achieve statistical power, and it ignores the uncertainty in the estimate. The original paper [20] used the first option; we are currently developing a hierarchical model to perform

differential splicing analyses while considering the uncertainty and heterogeneity within cluster.

2. As opposed to gene-level expression quantification, splicing quantification introduces a substantial uncertainty when there are very limited reads observed as often in scRNA-seq data. Therefore, it is not sensible to directly use the events-by-cell matrix of the mean estimates from BRIE in downstream analyses. A large fraction of splicing events has no or very few reads, namely the quantification is mainly based on the imputation, whose distribution is usually broad. Therefore, a single mean value may not be representative enough to the estimated distribution. There are two options to use them. The first option is to filter some events in some cells and treat them as missing values. The filtering condition could be based on the posterior distribution itself (e.g., 95% confidence interval <0.25) to select only high-confidence events, or on coverage thresholds (e.g. more than 5 reads or 3 junction reads). Filters based on confidence intervals in particular can be very useful, as the imputations in some events are highly informative and should be used, even there are fewer than 3 reads. The second option is to use both mean and variance of the estimate. Though this option accounts for the uncertainty in the estimate, it is not compatible with standard gene-level analysis, and requires some skills in statistical modeling.

3. Multiplexed scRNA-seq experiments are increasingly popular for drop-seq, SMART-seq and other protocols, thanks to its lower cost, lower batch effects, and ability for doublets detection. However, BRIE is not able to demultiplex these samples, and it relies on demultiplexed cells, namely each cell has its only aligned reads. One option for demultiplex is using cardelino (https://github.com/PMBio/cardelino), which includes a pipeline for SMART-seq/SMART-seq2 data.

4. Many software packages introduce new features or even new input/output formats during upgrading. Therefore, it is often important to specify the environment. Within the Python platform, many tools developed in Python2 may fail to run in Python3 environment. BRIE developers try their best to make it compatible with both Python2 and Python3. Should BRIE fail in one environment, e.g., Python3, the simplest strategy is to try the other environment, which can be very easily set up via the conda environment (https://conda.io/docs/user-guide/tasks/manage-environments.html).

## References

1. Grün D, van Oudenaarden A (2015) Design and analysis of single-cell sequencing experiments. Cell 163:799–810

2. Grün D, Lyubimova A, Kester L et al (2015) Single-cell messenger RNA sequencing reveals rare intestinal cell types. Nature 525:251–255

3. Gaublomme JT, Yosef N, Lee Y et al (2015) Single-cell genomics unveils critical regulators of Th17 cell pathogenicity. Cell 163:1400–1412

4. Papalexi E, Satija R (2018) Single-cell RNA sequencing to explore immune cell heterogeneity. Nat Rev Immunol 18:35

5. Scialdone A, Tanaka Y, Jawaid W et al (2016) Resolving early mesoderm diversification through single-cell expression profiling. Nature 535:289–293. https://doi.org/10.1038/nature18633

6. Wagner DE, Weinreb C, Collins ZM et al (2018) Single-cell mapping of gene expression landscapes and lineage in the zebrafish embryo. Science 80:eaar4362

7. Stubbington MJT, Lönnberg T, Proserpio V et al (2016) T cell fate and clonality inference from single-cell transcriptomes. Nat Methods 13:329

8. Lönnberg T, Svensson V, James KR et al (2017) Single-cell RNA-seq and computational analysis using temporal mixture modelling resolves Th1/Tfh fate bifurcation in malaria. Sci Immunol 2(9):eaal2192

9. Patel AP, Tirosh I, Trombetta JJ et al (2014) Single-cell RNA-seq highlights intratumoral heterogeneity in primary glioblastoma. Science 344:1396–1401

10. Tirosh I, Izar B, Prakadan SM et al (2016) Dissecting the multicellular exosystem of metastatic melanoma by single-cell RNA-seq. Science 352:189–196. https://doi.org/10.1126/science.aad0501.Dissecting

11. Wang ET, Sandberg R, Luo S et al (2008) Alternative isoform regulation in human tissue transcriptomes. Nature 456:470–476

12. Baralle FE, Giudice J (2017) Alternative splicing as a regulator of development and tissue identity. Nat Rev Mol Cell Biol 18:437

13. Dillman AA, Hauser DN, Gibbs JR et al (2013) mRNA expression, splicing and editing in the embryonic and adult mouse cerebral cortex. Nat Neurosci 16:499

14. Scotti MM, Swanson MS (2016) RNA mis-splicing in disease. Nat Rev Genet 17:19

15. Ziegenhain C, Vieth B, Parekh S et al (2017) Comparative analysis of single-cell RNA sequencing methods. Mol Cell 65:631–643

16. Faigenbloom L, Rubinstein ND, Kloog Y et al (2015) Regulation of alternative splicing at the single-cell level. Mol Syst Biol 11:845

17. Song Y, Botvinnik OB, Lovci MT et al (2017) Single-cell alternative splicing analysis with expedition reveals splicing dynamics during neuron differentiation. Mol Cell 67:148–161

18. La Manno G, Soldatov R, Hochgerner H et al (2018) RNA velocity of single cells. Nature 560.7719:494

19. Linker SM, Urban L, Clark S et al (2018) Combined single cell profiling of expression and DNA methylation reveals splicing regulation and heterogeneity. bioRxiv:328138

20. Huang Y, Sanguinetti G (2017) BRIE: transcriptome-wide splicing quantification in single cells. Genome Biol 18:123. https://doi.org/10.1101/098517

21. Huang Y, Sanguinetti G (2016) Statistical modeling of isoform splicing dynamics from RNA-seq time series data. Bioinformatics 32:2965–2972

22. Liu P, Sanalkumar R, Bresnick EH et al (2016) Integrative analysis with ChIP-seq advances the limits of transcript quantification from RNA-seq. Genome Res 26:1124–1133

23. Xiong HY, Alipanahi B, Lee LJ et al (2015) The human splicing code reveals new insights into the genetic determinants of disease. Science 1254806:347

24. Katz Y, Wang ET, Airoldi EM, Burge CB (2010) Analysis and design of RNA sequencing experiments for identifying isoform regulation. Nat Methods 7:1009–1015

25. Trapnell C, Williams BA, Pertea G et al (2010) Transcript assembly and quantification by RNA-Seq reveals unannotated transcripts and isoform switching during cell differentiation. Nat Biotechnol 28:511–515

26. Bray NL, Pimentel H, Melsted P, Pachter L (2016) Near-optimal probabilistic RNA-seq quantification. Nat Biotechnol 34:525

27. Kim D, Langmead B, Salzberg SL (2015) HISAT: a fast spliced aligner with low memory requirements. Nat Methods 12:357

28. Dobin A, Davis CA, Schlesinger F et al (2013) STAR: ultrafast universal RNA-seq aligner. Bioinformatics 29:15–21

29. Katz Y, Wang ET, Silterra J et al (2015) Quantitative visualization of alternative exon expression from RNA-seq data. Bioinformatics 31:2400–2402

# Chapter 13

## Preprocessing and Computational Analysis of Single-Cell Epigenomic Datasets

### Caleb Lareau, Divy Kangeyan, and Martin J. Aryee

### Abstract

Recent technological developments have enabled the characterization of the epigenetic landscape of single cells across a range of tissues in normal and diseased states and under various biological and chemical perturbations. While analysis of these profiles resembles methods from single-cell transcriptomic studies, unique challenges are associated with bioinformatics processing of single-cell epigenetic data, including a much larger (10–1,000×) feature set and significantly greater sparsity, requiring customized solutions. Here, we discuss the essentials of the computational methodology required for analyzing common single-cell epigenomic measurements for DNA methylation using bisulfite sequencing and open chromatin using ATAC-Seq.

**Key words** Epigenetics, Bioinformatics, Single-cell, DNA methylation, Bisulfite sequencing, ATAC-seq

## 1 Introduction

Single-cell epigenetic data analysis pipelines are designed to resolve differences between single cells in heterogeneous populations, including variability between and within classically defined cell types in DNA methylation using single-cell DNA methylation (scDNAm) profiles (Fig. 1) and chromatin accessibility using single-cell ATAC-Seq (scATAC-seq) profiles (Fig. 2). Axes of variation driving cell-to-cell epigenomic differences include cell-type, cell-state, and cell-cycle, similar to transcriptomic differences.

Figure 3 outlines the most common technology platforms associated with generating single-cell epigenomic data. While downstream analyses for each platform are largely the same, technology-specific bioinformatics solutions at the front-end of analyses are often required. For example, demultiplexing individual cells from a pool often requires custom scripts to efficiently parse out barcode sequences.

Guo-Cheng Yuan (ed.), *Computational Methods for Single-Cell Data Analysis*, Methods in Molecular Biology, vol. 1935, https://doi.org/10.1007/978-1-4939-9057-3_13, © Springer Science+Business Media, LLC, part of Springer Nature 2019

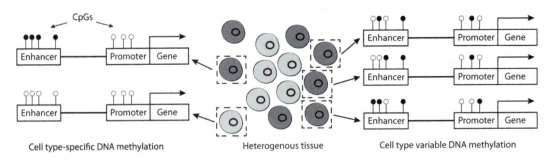

**Fig. 1** Overview of single-cell DNA methylation data. For a heterogeneous group of cells (middle), variability in CpG methylation can occur both between classically defined cell types (left) and within cell types (right)

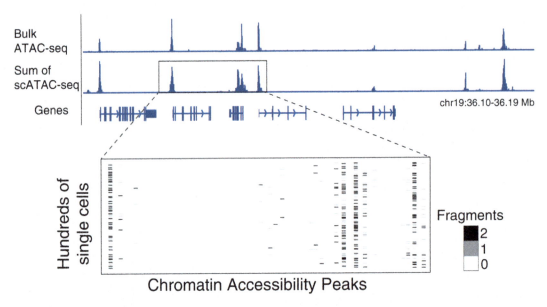

**Fig. 2** Overview of single-cell ATAC data. The sum of single cells' chromatin accessibility profiles resemble that of a bulk experiment (top) though each cell has a varied open chromatin epigenome. For a diploid organism, the number of accessible chromatin counts does not generally exceed 2

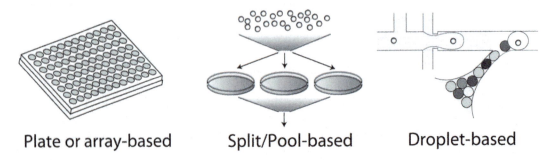

**Fig. 3** Modes of single-cell epigenomic assays. For scATAC and DNA methylation, popular approaches include array and split-pool-based though advances in microfluidic technologies will be increasingly used

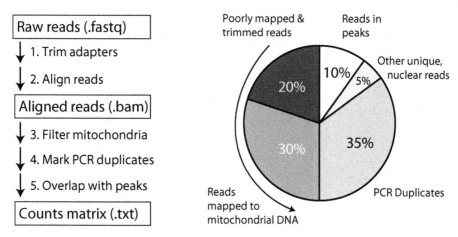

**Fig. 4** Overview of scATAC-seq data processing. Steps associated with data processing are shown on the left while a representative data allocation for a given cell is shown on the right. Roughly only about 10% of the raw data for a given scATAC-seq cell is used in downstream analyses, such as cell clustering

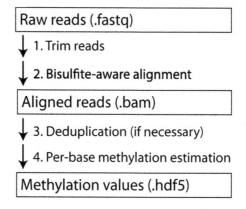

**Fig. 5** Overview of single-cell DNA methylation data processing. Steps associated with data processing are shown on the left

Here, we describe the essentials of processing single-cell epigenomic assays, starting with raw data preprocessing which typically involves demultiplexing pooled library FASTQ read files, the per-cell quantification of epigenomic state, and the collation of data from multiple cells into a single data structure (Figs. 4 and 5). Due to the large fraction of unobserved loci in single-cell epigenomic data, further downstream analysis is typically preceded by a step that aggregates data across multiple sites to reduce sparsity. This might involve biologically-motivated approaches such as averaging DNA methylation values across predefined sets of promoters or enhancers, or computing transcription factor accessibility metrics by combining scATAC-seq data from all sites for a given factor (Fig. 6). Notably, only a fraction of the full data is fully utilized in downstream applications, such as ~10% of reads for a given cell from scATAC-seq library preparation (Fig. 4).

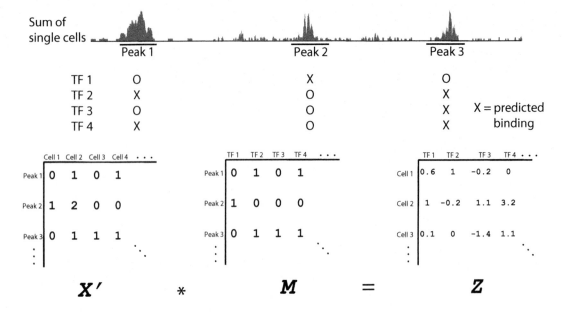

**Fig. 6** Biologically-motivated scATAC-seq dimensionality reduction. A binary matrix of motifs by peaks is multiplied by an integer-value matrix of peaks by cells to yield reduced set of features (real valued motif deviation scores) per cell. These can be utilized for downstream analyses, including visualization with t-SNE, and cell type identification

## 2 Materials

The protocols described here rely on free, open-source bioinformatics tools. All tools can be installed on a Mac or Linux computer. The single-cell ATAC-seq protocol uses:

1. chromVAR (https://bioconductor.org/packages/release/bioc/html/chromVAR.html) [1].

2. Picard tools (https://broadinstitute.github.io/picard/).

The single-cell DNA methylation protocol uses:

3. Bismark (https://www.bioinformatics.babraham.ac.uk/projects/bismark/) [2].

Shared tools useful for both protocols include the following software:

4. bcl2fastq (https://support.illumina.com/sequencing/sequencing_software/bcl2fastq-conversion-software.html).

5. Bowtie2 (http://bowtie-bio.sourceforge.net/bowtie2/) [3].

6. TrimGalore (https://www.bioinformatics.babraham.ac.uk/projects/trim_galore/).

7. R (https://cran.r-project.org).

8. Bioconductor (https://www.bioconductor.org) [4].

9. bedtools (http://bedtools.readthedocs.io/en/latest/index.html) [5].

10. bedgraphToBigwig      (http://hgdownload.soe.ucsc.edu/admin/exe/).

---

## 3 Methods

The protocols described below share the first step (Subheading 3.1) and are then described separately for single-cell DNA ATAC-Seq (Subheading 3.2) and single-cell DNA methylation (Subheading 3.3). Graphical overviews of these steps for each method are shown in Figs. 4 and 5.

### 3.1 Demultiplexing Single-Cell Libraries

1. Single-cell experimental protocols typically involve combining individually barcoded single-cell libraries into pools for high-throughput sequencing. The first step in preprocessing is thus to demultiplex reads into per-cell FASTQ files. The pooling is often done in a two-step process (*see* **Note 1**). First, a set of single cells are individually tagged with barcodes and batched together into a Level 1 pool. This step often involves custom barcodes. Next, such a pool can then tagged with an additional barcode and pooled with others creating a Level 2 pool. This second level of pooling typically uses standard Illumina sample barcodes. The two-step pooling procedure thus requires two sequential demultiplexing steps.

2. The Illumina barcoded Level 2 pools can be demultiplexed using the standard Illumina bcl2fastq tool to generate individual fastq files for each Level 1 pool, e.g.:

```
bcl2fastq --runfolder-dir NextSeq_Data --output-dir
fastq_directory Samplesheet.csv
```

3. Next, we demultiplex the Level 1 pool fastqs to generate a fastq (pair) for each single cell. This step typically requires protocol-specific tools since the exact single-cell barcode structure used varies from protocol to protocol. For data generated from the Fluidigm C1 array, these indices are typically codes in the P5 and P7 adaptor sequences. Otherwise, adaptor sequences have to be parsed out in-line with custom scripts.

### 3.2 scATAC-Seq

1. Trim adapter sequences from fastq files using tools such as TrimGalore (*see* **Note 2**).

```
trim_galore --paired read1.fastq.gz read2.fastq.gz
```

2. Align trimmed reads with `bowtie2` and pipe to `samtools` to create a sorted `.bam` file (*see* **Note 3**).

```
bowtie2 -X 2000 -1 cellA_1.trim.fastq.gz -2
cell_A2.trim.fastq.gz -x /path/to/bowtie2/reference |
samtools view -bS - | samtools sort - -o cellA.st.bam
samtools index cellA.st.bam
```

3. Filter reads for unique alignments `Picard` tools.

```
java -jar MarkDuplicates.jar I=cellA.st.bam
O=cellA.dedup.bam M=cellA.marked_rep
ort.txt VALIDATION_STRINGENCY=SILENT REMOVE_DUPLICATES=true
```

4. Shift aligned reads based on Tn5 activity (*see* **Note 4**).

```
bedtools bamtobed -i reads.bam -bedpe | awk -v OFS="\t"
'{if($9=="+"){print $1,$2+4,$6+4}else if($9=="-"){print
$1,$2-5,$6-5}}' > cellA_fragments.bed
```

5. Remove reads overlapping known blacklist regions (*see* **Note 5**).

```
bedtools subtract -a cellA_fragments.bed -b blacklist.bed >
cellA_filt_frags.bed
```

6. Call accessibility peaks using `MACS2` [6] on shifted reads over the union of cell fragments (*see* **Notes 6** and **7**).

```
macs2 call peak -t allCells_fragments.bed --nomodel --shift
-100 --extsize 200 --keep-dup all --call-summits -n combined-
Cells
```

7. Assemble peak by cell counts matrix using an interactive R session with `chromVAR` (*see* **Note 8**).

```
library(chromVAR)
library(motifmatchr)
library(BSgenome.Hsapiens.UCSC.hg19)  # change based on
reference genome
library(SummarizedExperiment)
library(chromVARmotifs)
peakfile <- "data/peaks.bed"
peaks <- getPeaks(peakfile)
# Import fragments from per-cell
bedfiles <- list.files("data/cell_fragments", full.names =
TRUE)
raw_counts <- getCounts(bamfiles, peaks, by_rg = FALSE,
format = "bed",
colData = DataFrame(source = bedfiles))
```

8. Perform scATAC-seq-specific dimensionality reduction (*see* **Notes 9** and **10**).

```
# Initialize parallel processing
library(BiocParallel)
register(MulticoreParam(2)) # adjust according to your
machine

# Get GC content/peak; get motifs from chromVARmotifs
package
counts <- addGCBias(counts, genome =
BSgenome.Hsapiens.UCSC.hg19)
data("human_pwms_v2") # also mouse_pwms_v2
motif_ix <- matchMotifs(human_pwms_v2, counts, genome =
BSgenome.Hsapiens.UCSC.hg19)
dim(motif_ix)

# Compute deviations; typically
most time consuming step
dev <- computeDeviations(object = counts, annotations =
motif_ix)
```

9. Visualize scATAC cells in a two dimensional space by supplying deviations scores to the t-distributed Stochastic Neighbor Embedding algorithm (*see* **Note 11**).

*3.3  scMethylation*

1. Trim sequencing adapter sequences from fastq files. It is also advisable to additionally trim of poor quality bases at the ends of reads that can lead to alignment errors and/or incorrect methylation calls. For example, to perform both quality and adapter trimming in a single step one can use `TrimGalore` (*see* **Note 12** for RRBS libraries):

```
trim_galore --paired read1.fastq.gz read2.fastq.gz
```

2. Align trimmed reads with a bisulfite-aware aligner (*see* **Note 13**).

```
bismark --genome Bisulfite_Genomes/grch38  -1 fastq1.trim.
fastq.gz -2 fastq2.trim.fastq.gz
```

3. Remove PCR duplicates.

This step should be performed only for whole-genome bisulfite sequencing (WGBS) libraries. It is not advised for reduced representation (RRBS) or targeted capture libraries (*see* **Note 14**).

```
deduplicate_bismark --bam cell_1.bam
```

4. Quantify methylation for each cell at each CpG position (*see* **Note 15**).

Per-CpG methylation can be estimated as the fraction of reads representing methylated cytosines over the total number of reads at the position. We report the number of methylated reads (M), unmethylated reads (U) and the ratio M/(M+U).

```
bismark_methylation_extractor --gzip --bedGraph --
genome_folder bismark_index cell_1.bam
```

By default, this command will omit CpGs with no coverage. Forcing the inclusion of all CpGs (even those with zero coverage) by adding the --cytosine_report option can be helpful for simplifying and speeding up the downstream step of combining data across cells (*see* **Note 16**).

5. (Optional) Generate a bigwig track for visualization in a genome browser.

   Individual cell methylation values can be visualized in IGV or another browser using a bigwig track file (*see* **Note 16**).

```
gunzip cell_1.bedGraph.gz
bedtools sort -i cell_1.bedGraph > cell_1.sorted.bedGraph
bedGraphToBigWig cell_1.sorted.bedGraph
grch38_chrom_sizes.txt cell_1.bw
```

6. Collate data from individual cells.

   Workflow **steps 1–5** are performed on a per-cell basis and can therefore be parallelized. For downstream analysis, it is often convenient to have data from all cells represented in a single data structure, such as a matrix of methylation estimates with one row per CpG (or genomic region), and one column per cell. This collation can be performed in R/Bioconductor using the bsseq (https://bioconductor.org/packages/release/bioc/html/bsseq.html) package (*see* **Notes 18** and **19** for a basic example script). As an alternative collation methods for users who prefer not to use R, one can also simply paste together the methylation estimate columns (column 4) from the individual cell *.cov.gz files from **step 4** (e.g., using the Unix paste command*)*. In order to allow this one needs to ensure that all files have exactly the same CpG order (*see* **Note 17**).

7. Aggregate data across CpGs.

   Unlike transcriptomic assays, single-cell DNA-based assays like scDNAm do not benefit from the amplification inherent in transcription. As a result the number of template molecules can be as low as one, leading to a very large fraction (often >90%) of missing data. As a result, analyses of individual sites (e.g., specific gene promoters) is challenging as a given gene would only be observed in a small fraction of cells. A common approach to dealing with this sparsity involves averaging methylation values across all CpGs in a predefined feature set. These feature sets consist of genomic regions relevant to the downstream questions of interest but may include, for example, gene promoters/enhancers for each Gene Ontology category, or various classes of repetitive elements. This process will transform the highly sparse matrix from **step 6** (one row per CpG, one column per cell) to a smaller, denser matrix with one row per feature set and one column per cell. This dense feature set matrix is a starting point for analyses that is amenable to standard approaches such as PCA, clustering or identification of differentially methylated feature sets across.

---

## 4    Notes

1. The number of cells per pool is limited either by the technology platform (the Fluidigm C1 array, for example, processes cells in batches of 96 cells), or by the need to generate sufficient sequencing depth per cell.

   Most single-cell ATAC-seq libraries are processed from paired-end, dual-index sequencing. Previous studies using the Fluidigm C1 array [7] and split-pool [8] index cells using unique combinations of the indices from the sequencing technology.

2. The two common Nextera adapters used in ATAC-seq and bisulfite sequencing are TCGTCGGCAGCGTC and TGGTAGA-GAGGGTG. These will be the most typical sequences that must be trimmed before alignment.

3. The --X 2000 flag is used to increase alignment of reads spanning 2–5 nucleosomes (the width of one nucleosome is ~147 bp) which have been reliably detected from ATAC-seq data. The default option in bowtie2 is 500. Additionally, these steps can benefit from multiple available cores through parallel execution using the -p # flag for bowtie2 and the -@ # flag in samtools where # is an integer (2 for standard laptop computers or as high as 32 on a high-performance computing environment).

4. After processing reads for PCR duplicates and proper pairs, ATAC-seq and scATAC-seq aligned read coordinates are typically offset by +4 and the minus strand aligning reads by −5 bp to represent the center of the transposon binding event. This bias was first reported in the original ATAC-seq paper [9]. Depending on the downstream applications of the scATAC data, invoking this option later may be desirable.

5. The standard ENCODE blacklist for hg19 can be found here: http://hgdownload.cse.ucsc.edu/goldenPath/hg19/encodeDCC/wgEncodeMapability/. An additional custom blacklist for ATAC-seq derived from mitochondrial reads mapped to the nuclear genome can be found here: https://github.com/buenrostrolab/mitoblacklist.

6. The `--shift -100 --extsize 200` option centers a 200 bp window where the Tn5 binds, yielding more accurate peak accessibility coordinates. The `--nomodel` flag bypasses the read shifting model that was designed for ChIP-seq data. Some users also invoke the `--nolambda` flag to.

7. A strategy for integrating peak calls across multiple cell types or batches for analysis is discussed here: http://bioconductor. org/packages/devel/bioc/vignettes/DiffBind/inst/doc/ DiffBind.pdf. In short, using called accessibility summits and then padding coordinates by fixed amounts can lead to fixed-width peaks and an inclusive feature set, both of which have many desirable properties.

8. When importing data into `chromVAR`, it is often more convenient to have all the single-cell reads contained in a single `.bam` file rather than split per-file, especially if the cell count exceeds 100. To achieve this, you can add a read group (RG) ID at the `bowtie2` alignment step:

```
bowtie2 -1 cellA_1.trim.fastq.gz -2 cell_A2.trim.fastq.gz
--rg-id cellA ...
```

9. Intuitively, this approach will compute bias-corrected deviation scores for each of the $K$ motifs and a set of $S$ samples (scATAC-seq profiles) with $P$ peaks. A positive deviation score can be interpreted as the specific cell has more accessible chromatin at the specific genomic feature than expected by chance. This procedure in `chromVAR` also requires binarized matrix $M$ (dimension $P$ by $K$) where $m_{i,k}$ is 1 if annotation $k$ (such as a predicted transcription factor binding site) is present in peak $i$ and 0 otherwise.

Using the matrix of fragment counts in peaks $X$, where $x_{i,j}$ represents the number of fragments from peak $i$ in sample $j$, a matrix multiplication $X^T \cdot M$ yields the total number of fragments weighted by the presence of a predicted TF binding site. To compute a raw weighted accessibility deviation, we compute the expected number of fragments per peak per sample in $E$, where $e_{i,j}$ is computed as the proportion of all fragments across all samples mapping to the specific peak multiplied by the total number of fragments in peaks for that sample:

$$e_{i,j} = \frac{\sum_j x_{i,j}}{\sum_j \sum_i x_{i,j}} \Sigma_i x_{i,j}$$

Analogously, $X^T \cdot E$ yields the expected number of fragments weighted by the fine mapped variant posterior probabilities for $S$ samples (rows) and $K$ factors (columns). Using the $M, X,$

and $E$ matrices, we then compute the raw weighted accessibility deviation matrix $\Upsilon$ for each sample $j$ and trait $k$ ($y_{j,k}$) as follows:

$$y_{j,k} = \frac{\sum\limits_{i=1}^{P} x_{i,j}m_{i,k} - \sum\limits_{i=1}^{P} e_{i,j}m_{i,k}}{\sum\limits_{i=1}^{P} e_{i,j}m_{i,k}}$$

To correct for technical confounders present in assays (differential PCR amplification or variable Tn5 tagmentation conditions), chromVAR generates a background set of peaks intrinsic to the set of epigenetic data examined. The background peak sampling procedure has been described in depth elsewhere. [7,1] Ultimately, this procedure yields a matrix $B^{(b)}$. This matrix encodes this background peak mapping where $b_{i,j}^{(b)}$ is 1 if peak $i$ has peak $j$ as its background peak in the $b$ background set ($b \in \{1, 2, ..., 50\}$) and 0 otherwise. The matrices $B^{(b)} \cdot X$ and $B^{(b)} \cdot E$ thus give an intermediate for the observed and expected counts also of dimension $P$ by $S$. For each background set $b$, sample $j$, and trait $k$, the elements $y_{j,k}^{(b)}$ of the background weighted accessibility deviations matrix $\Upsilon^{(b)}$ are computed as follows:

$$y_{j,k}^{(b)} = \frac{\sum\limits_{i=1}^{P} \left(B^{(b)} \cdot X\right)_{i,k} m_{i,k} - \sum\limits_{i=1}^{P} \left(B^{(b)} \cdot E\right)_{i,k} m_{i,k}}{\sum\limits_{i=1}^{P} \left(B^{(b)} \cdot E\right)_{i,k} m_{i,k}}$$

After the background deviations are computed over the 50 sets, the bias-corrected matrix $Z$ for sample $j$ and trait $k$ ($z_{j,k}$) can be computed as follows:

$$z_{j,k} = \frac{y_{j,k} - \text{mean}\left(y_{j,k}^{(b)}\right)}{\text{sd}\left(y_{j,k}^{(b)}\right)}$$

where the mean and variance of $y_{j,k}^{(b)}$ is taken over all values of $b$ ($b \in \{1, 2, ..., 50\}$). This implementation uses computationally efficient matrix operations for each step and can compute pairwise trait-cell type enrichments in ~1 min on a standard laptop computer.

10. A recent extension of the `chromVAR` methodology (called `g-chromVAR`) to handle weighted or uncertain motif annotations specifically designed for variants nominated by genome-wide association studies has recently been described [10].

11. A sample `Rscript` used to perform **steps 7–9** from `.bam` files (the most common workflow) is shown here:

```
library(chromVAR)
library(motifmatchr)
library(BSgenome.Hsapiens.UCSC.hg19) # change based on
reference genome
library(SummarizedExperiment)
library(chromVARmotifs)

# Initialize parallel processing
library(BiocParallel)
register(MulticoreParam(2)) # adjust according to your
machine

# Import/filter data (replace with appropriate file paths
peakfile <- "data/peaks.bed"
peaks <- getPeaks(peakfile)
bamfiles <- list.files("data/bams", full.names = TRUE)
raw_counts <- getCounts(bamfiles, peaks, paired =  TRUE,
by_rg = FALSE,
format = "bam", colData = DataFrame(source = bamfiles))

# Filter low quality samples and peaks
counts_filtered <- filterSamples(raw_counts, min_depth =
500,
                                 min_in_peaks = 0.15, shiny =
FALSE)
counts <- filterPeaks(counts_filtered)

# Get GC content/peak; get motifs from chromVARmotifs
package; find kmers
counts <- addGCBias(counts, genome = BSgenome.Hsapiens.
UCSC.hg19)
data("human_pwms_v2") # also mouse_pwms_v2
motif_ix <- matchMotifs(human_pwms_v2, counts, genome =
BSgenome.Hsapiens.UCSC.hg19)
dim(motif_ix)

# Compute deviations; typically most time consuming step
dev <- computeDeviations(object = counts, annotations =
motif_ix)

# Find variable motifs
variabilityAll <- computeVariability(dev)
plotVariability(variabilityAll, use_plotly = FALSE)

# Visualize single cells with a tSNE
tsne_results <- deviationsTsne(dev, threshold = 1.5, per-
plexity = 10,
```

```
                                              shiny = FALSE)
tsne_plots <- plotDeviationsTsne(dev, tsne_results, anno-
tation = "CTCF",
                                       sample_column = "source",
shiny = FALSE)
```

12. The MspI-digestion Reduced Representation Bisulfite Sequencing (RRBS) protocol involves an end-repair step that introduces an unmethylated cytosine at the end of fragments. This cytosine should not be used for methylation calling and can be trimmed by removing 2 bases at the end of reads using the --rrbs option of TrimGalore:

```
trim_galore --rrbs --paired read1.fastq.gz read2.fastq.gz
```

13. The alignment step required a preindexed bisulfite converted genome. This can be generated using the bismark_genome_preparation tool. The single required argument points to a directory that contains the reference genome FASTA file (s).

```
bismark_genome_preparation /path/to/genomes/grch38/
```

14. Deduplication is typically performed by retaining only one read (or read pair) for each unique mapping start position. This is done under the assumption that the number of potential molecules to sample from is large (on the order of the number of base pairs in the genome) compared to the number of sequenced reads, and thus reads with the same start position are more likely to represent PCR duplicates than to have arisen from different pre-amplification DNA fragments. However, since reduced representation bisulfite sequencing (RRBS) libraries and capture/hybrid selection bisulfite libraries only represent a small portion of the genome this assumption does not hold.

15. Adding the --CX_context option to bismark_methylation_extractor will quantify methylation at all cytosines (on both strands), as opposed to only those in the CpG context. Note that this will generate an output file with over a billion lines for a human genome.

16. A similar set of commands can be run for visualizing scATAC data in the .bigwig format. Similar to the single-cell methylation data, these accessibility profiles should most likely be pooled across cells. Note that bedGraphToBigWig takes an input file that specifies the length of each chromosome. This can be generated using samtools as follows:

```
samtools faidx grch38.fa
cut -f1,2 grch38.fa.fai > grch38_chrom_sizes.txt
```

17. Current single-cell methylation assays are able to capture only a small fraction of the ~28 million CpGs in the genome. As a result the CpG coverage and methylation output of bismark_methylation_extractor will contain a different set of cytosines for each cell. Excluding the uninformative cytosines with zero coverage saves disk space, but complicates the step of collating data from multiple cells into a single table containing the union of all observed CpGs across. The `bismark_methylation_extractor --cytosine_report` option will force it to output all CpGs, thus creating a common set of rows for all cells. This allows collation to be done simply by pasting together the methylation estimate columns from each cell's coverage file.

18. The `R/Bioconductor` script below is a basic template for collating the per-cell cytosine coverage outputs from `bismark_methylation_extractor`:

```
# Replace this with the location/names of the output from
bismark_methylation_extractor
covgz_files <- c("cell_1.bismark.cov.gz",
                 "cell_2.bismark.cov.gz",
                 "cell_3.bismark.cov.gz")

library(Biostrings)
library(bsseq)
library(BSgenome.Hsapiens.NCBI.GRCh38)
library(readr)
library(HDF5Array)

getMethCov <- function(covgz_file, gr) {
  tab <- read_tsv(covgz_file,
            col_types = "ciidii",
            col_names=c("chr", "pos", "pos2", "meth_per-
cent", "m_count", "u_count"))
  tab_gr <- GRanges(tab$chr, IRanges(tab$pos, tab$pos))
  m <- rep(0, length(gr))
  cov <- rep(0, length(gr))
  ovl <- suppressWarnings(findOverlaps(tab_gr, gr))
  m[subjectHits(ovl)]   <- tab$m_count[queryHits(ovl)]
  cov[subjectHits(ovl)] <- tab$m_count[queryHits(ovl)] +
tab$u_count[queryHits(ovl)]
  return(list(m=m, cov=as.integer(cov)))
}

# Set up genome-wide CpG GRanges
```

```
# On the plus strand we keep the left-most position of the
match
# On the minus strand we keep the right-most position of
the match
cpg_gr <- DNAString("CG")
cpg_gr <- vmatchPattern(cpg_gr, BSgenome.Hsapiens.NCBI.
GRCh38)
cpg_gr <- keepStandardChromosomes(cpg_gr, pruning.mode="-
coarse")
s <- start(cpg_gr)
e <- end(cpg_gr)
plus_idx <- as.logical(strand(cpg_gr)=="+")
minus_idx <- as.logical(strand(cpg_gr)=="-")
e[plus_idx] <- s[plus_idx]  # Plus strand
s[minus_idx] <- e[minus_idx] # Minus strand
start(cpg_gr) <- s
end(cpg_gr) <- e

# Get methylation, coverage matrices and store on disk as
HDF5Arrays
hdf5_m <- list()
hdf5_cov <- list()
for(covgz_file in covgz_files) {
  tmp <- getMethCov(covgz_file, cpg_gr)
  m <- tmp$m
  cov <- tmp$cov
  samplename <- basename(covgz_file)
  samplename <- gsub(samplename, pattern="_PE_report.txt",
replacement="")
  hdf5_file <- paste0(samplename, ".hdf5")
  if(file.exists(hdf5_file)) file.remove(hdf5_file)
    hdf5_m[[samplename]] <- writeHDF5Array(matrix(m),
name="m", file=hdf5_file)
    hdf5_cov[[samplename]] <- writeHDF5Array(matrix(cov),
name="cov", file=hdf5_file)
}
M <- do.call("cbind", hdf5_m)
Cov <- do.call("cbind", hdf5_cov)

# Create a bsseq object ready for downstream analysis
bs <- BSseq(gr=cpg_gr, M=M, Cov=Cov)
```

19. We provide sample code for running specific downstream parts of these described methods here: https://github.com/aryeelab/mmb-scEpigenomics

## Acknowledgments

We are grateful to Jason Buenrostro for useful feedback in the discussion of the scATAC-seq computational analyses.

## References

1. Schep AN, Wu B, Buenrostro JD, Greenleaf WJ (2017) chromVAR: inferring transcription-factor-associated accessibility from single-cell epigenomic data. Nat Methods 14 (10):975–978. https://doi.org/10.1038/nmeth.4401

2. Krueger F, Andrews SR (2011) Bismark: a flexible aligner and methylation caller for Bisulfite-Seq applications. Bioinformatics 27 (11):1571–1572. https://doi.org/10.1093/bioinformatics/btr167

3. Langmead B, Salzberg SL (2012) Fast gapped-read alignment with Bowtie 2. Nat Methods 9 (4):357–359. https://doi.org/10.1038/nmeth.1923

4. Huber W, Carey VJ, Gentleman R, Anders S, Carlson M, Carvalho BS, Bravo HC, Davis S, Gatto L, Girke T, Gottardo R, Hahne F, Hansen KD, Irizarry RA, Lawrence M, Love MI, MacDonald J, Obenchain V, Oles AK, Pages H, Reyes A, Shannon P, Smyth GK, Tenenbaum D, Waldron L, Morgan M (2015) Orchestrating high-throughput genomic analysis with bioconductor. Nat Methods 12(2):115–121. https://doi.org/10.1038/nmeth.3252

5. Quinlan AR, Hall IM (2010) BEDTools: a flexible suite of utilities for comparing genomic features. Bioinformatics 26(6):841–842. https://doi.org/10.1093/bioinformatics/btq033

6. Zhang Y, Liu T, Meyer CA, Eeckhoute J, Johnson DS, Bernstein BE, Nusbaum C, Myers RM, Brown M, Li W, Liu XS (2008) Model-based analysis of ChIP-Seq (MACS). Genome Biol 9(9):R137. https://doi.org/10.1186/gb-2008-9-9-r137

7. Buenrostro JD, Wu B, Litzenburger UM, Ruff D, Gonzales ML, Snyder MP, Chang HY, Greenleaf WJ (2015) Single-cell chromatin accessibility reveals principles of regulatory variation. Nature 523(7561):486–490. https://doi.org/10.1038/nature14590

8. Cusanovich DA, Daza R, Adey A, Pliner HA, Christiansen L, Gunderson KL, Steemers FJ, Trapnell C, Shendure J (2015) Multiplex single cell profiling of chromatin accessibility by combinatorial cellular indexing. Science 348 (6237):910–914. https://doi.org/10.1126/science.aab1601

9. Buenrostro JD, Giresi PG, Zaba LC, Chang HY, Greenleaf WJ (2013) Transposition of native chromatin for fast and sensitive epigenomic profiling of open chromatin, DNA-binding proteins and nucleosome position. Nat Methods 10(12):1213–1218. https://doi.org/10.1038/nmeth.2688

10. Lareau CA, Ulirsch JC, Bao EL, Ludwig LS, Guo MH, Benner C, Satpathy AT, Salem R, Hirschhorn JN, Finucane HK, Aryee MJ, Buenrostro JD, Sankaran VG (2018) Interrogation of human hematopoiesis at single-cell and single-variant resolution. bioRxiv. https://doi.org/10.1101/255224

# Chapter 14

## Experimental and Computational Approaches for Single-Cell Enhancer Perturbation Assay

### Shiqi Xie and Gary C. Hon

### Abstract

Transcriptional enhancers drive cell-type-specific gene expression patterns, and thus play key roles in development and disease. Large-scale consortia have extensively cataloged >one million putative enhancers encoded in the human genome. But few enhancers have been endogenously tested for function. For almost all enhancers, it remains unknown what genes they target and how much they contribute to target gene expression. We have previously developed a method called Mosaic-seq, which enables the high-throughput interrogation of enhancer activity by performing pooled CRISPRi-based epigenetic suppression of enhancers with a single-cell transcriptomic readout. Here, we describe an optimized version of this method, Mosaic-seq2. We have made several key improvements that have significantly simplified the library preparation process and increased the overall sensitivity and throughput of the method.

**Key words** Single-cell RNA-seq, Enhancer, CRISPRi, Single-cell perturbation

## 1 Introduction

Changes in cell state are often driven by changes in gene expression. However, we currently lack a predictive understanding of gene regulation, and this is a major impediment to a deterministic understanding of development and disease. Regulatory elements called transcriptional enhancers are the key drivers of tissue- and disease-specific gene expression, and major consortia have identified more than one million putative enhancers in the human genome [1, 2]. A unique feature of enhancers is their ability to activate genes from large distances, occasionally >1 megabase away [3]. This feature has significantly complicated the accurate identification of enhancer target genes. Correlation analyses have predicted enhancer target genes, but few have been experimentally validated. As a result, despite the many developmental and disease systems in which enhancers have been mapped, the vast majority of predicted enhancers have no known targets. This is a key obstacle to determining the biomedical roles of each enhancer.

Guo-Cheng Yuan (ed.), *Computational Methods for Single-Cell Data Analysis*, Methods in Molecular Biology, vol. 1935, https://doi.org/10.1007/978-1-4939-9057-3_14, © Springer Science+Business Media, LLC, part of Springer Nature 2019

We have recently developed Mosaic-seq, a highly multiplexed enhancer perturbation technology that assesses endogenous enhancer activities without the gene tagging or phenotypic selection [4]. By combining CRISPRi [5, 6] with single-cell RNA sequencing, Mosaic-seq simultaneously measures sgRNA perturbations and the transcriptomic outcomes of these perturbations in the same single cell. This strategy is analogous to those described in Perturb-seq, CRISP-seq, and CROP-seq [7–9]. Mosaic-seq can unbiasedly identify the target genes of any enhancer. Moreover, it also provides unique information about single-cell enhancer usage and combinatorial enhancer activity [4].

In this chapter, we will introduce Mosaic-seq2, with key changes to increase sensitivity and throughput (Fig. 1). First, we adapt the 10× Genomics single-cell RNA-seq platform, which enables the generation of 80,000 single-cell RNA-Seq libraries in a single run. Second, we incorporate the CROP-seq design for sgRNA expression, which enables direct detection of sgRNAs without the need to barcode [9]. This design dramatically simplifies the construction of sgRNA plasmid libraries, and also eliminates the shuffling of sgRNAs and barcodes due to retroviral recombination [10, 11]. Third, as in Perturb-seq, we implement an enrichment PCR step to increase the detection efficiency of the sgRNAs in each single cell [7]. These improvements significantly increase the throughput of Mosaic-seq, which now allows simultaneous perturbation of hundreds to thousands of enhancers.

Below, we will describe Mosaic-seq2 using K562 cells as an example. This chapter details (1) the design and construction of the sgRNA libraries, (2) lentiviral packaging and infection, (3) preparation of single-cell RNA-seq libraries, and (4) computational analysis. This protocol can be readily extended to other systems. Overall, single-cell perturbation approaches like Mosaic-seq2 will enable an understanding of how enhancers function in single cells and how they define a cell's transcriptional regulatory network.

## 2 Materials

### 2.1 Cell Culture

1. K562 cells and HEK293T cells (both from ATCC).
2. Phosphate-buffered saline (PBS), pH 7.4.
3. 0.25% Trypsin-EDTA.
4. Complete cell culture medium: Iscove's Modified Dulbecco's Medium (IMDM, for K562) or Dulbecco's Modified Eagle's Medium (DMEM, for HEK293T) with 10% FBS, 100 U/mL Penicillin-Streptomycin (Thermo Fisher Scientific).
5. 10 cm cell culture dishes.
6. Hemocytometer.

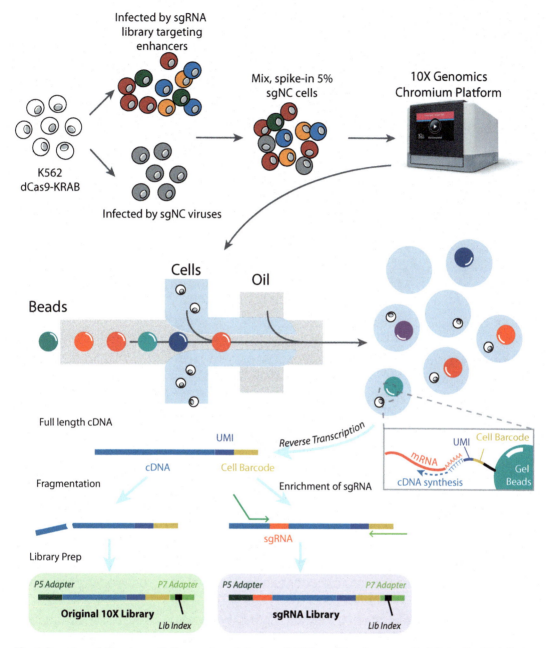

**Fig. 1** Overview of Mosaic-seq2. Preparation of single-cell RNA-seq libraries generally follows the $10\times$ library preparation procedures except that an sgRNA enrichment library is amplified from full length cDNA

***2.2   Plasmids and sgRNA Library Construction***

1. Lentivirus packaging plasmids: pMD2.G and psPAX2 (Addgene ID 12259 and 12260).

2. CROP-seq plasmid for sgRNA expression (Addgene ID 86708).

3. lenti-dCas9-KRAB plasmids (Addgene ID 89567).

4. sgRNA oligo library (Table 1).

**Table 1**
**List of oligos (*see* Note 1)**

| Name | Sequence |
|---|---|
| Oligo Amp Fwd | TAACTTGAAAGTATTTCGATTTCTTGGCTTTATATATCTTG TGGAAAGGACGAAACACCG |
| Oligo Amp Rev | ATTTTAACTTGCTAGGCCCTGCAGACATGGGTGATCCTCATGTTGGCC TAGCTCTAAAAC |
| sgRNA Oligo Pool | GTGGAAAGGACGAAACACCNNNNNNNNNNNNNNNNNNNNNNGTTTTAGAGC TAGGCCAACATGAGGATCAC |
| LKO1_5 primer | GACTATCATATGCTTACCGT |
| sgRNA_amp | TTGGCCTAGCTCTAAAAC |
| sgRNA-Lib staggered i5-1 | AATGATACGGCGACCACCGAGATCTACACTCTTTCCCTACACGACGCTC TTCCGATCTCTATCATATGCTTACCGT |
| sgRNA-Lib staggered i5-2 | AATGATACGGCGACCACCGAGATCTACACTCTTTCCCTACACGACGCTC TTCCGATCTACTATCATATGCTTACCGT |
| sgRNA-Lib staggered i5-3 | AATGATACGGCGACCACCGAGATCTACACTCTTTCCCTACACGACGCTC TTCCGATCTTACTATCATATGCTTACCGT |
| sgRNA-Lib staggered i5-4 | AATGATACGGCGACCACCGAGATCTACACTCTTTCCCTACACGACGCTC TTCCGATCTGTTCTATCATATGCTTACCGT |
| sgRNA-Lib staggered i5-5 | AATGATACGGCGACCACCGAGATCTACACTCTTTCCCTACACGACGCTC TTCCGATCTTGTACTATCATATGCTTACCGT |
| sgRNA-Lib i7-N720 | CAAGCAGAAGACGGCATACGAGATAGGCTCCGGTGACTGGAGTTCAGACG TGTGCTCTTCCGATCTTTGGCCTAGCTCTAAAAC |
| sgRNA-Lib i7-N721 | CAAGCAGAAGACGGCATACGAGATGCAGCGTAGTGACTGGAGTTCAGACG TGTGCTCTTCCGATCTTTGGCCTAGCTCTAAAAC |
| sgRNA-Lib i7-N722 | CAAGCAGAAGACGGCATACGAGATCTGCGCATGTGACTGGAGTTCAGACG TGTGCTCTTCCGATCTTTGGCCTAGCTCTAAAAC |
| sgRNA-Lib i7-N723 | CAAGCAGAAGACGGCATACGAGATGAGCGCTAGTGACTGGAGTTCAGACG TGTGCTCTTCCGATCTTTGGCCTAGCTCTAAAAC |
| sgRNA-Lib i7-N724 | CAAGCAGAAGACGGCATACGAGATCGCTCAGTGTGACTGGAGTTCAGACG TGTGCTCTTCCGATCTTTGGCCTAGCTCTAAAAC |
| sgRNA-Lib i7-N726 | CAAGCAGAAGACGGCATACGAGATGTCTTAGGGTGACTGGAGTTCAGACG TGTGCTCTTCCGATCTTTGGCCTAGCTCTAAAAC |
| SI-PCR primer | AATGATACGGCGACCACCGAGATCTACACTCTTTCCCTACACGACGCTC |
| 10×-sgRNA i7-N720 | CAAGCAGAAGACGGCATACGAGATAGGCTCCGGTGACTGGAGTTCAGACG TGTGCTCTTCCGATCTTGGAAAGGACGAAACACC |
| 10×-sgRNA i7-N721 | CAAGCAGAAGACGGCATACGAGATGCAGCGTAGTGACTGGAGTTCAGACG TGTGCTCTTCCGATCTTGGAAAGGACGAAACACC |
| 10×-sgRNA i7-N722 | CAAGCAGAAGACGGCATACGAGATCTGCGCATGTGACTGGAGTTCAGACG TGTGCTCTTCCGATCTTGGAAAGGACGAAACACC |

(continued)

**Table 1**
**(continued)**

| Name | Sequence |
|------|----------|
| 10×-sgRNA i7-N723 | CAAGCAGAAGACGGCATACGAGATGAGCGCTAGTGACTGGAGTTCAGACG TGTGCTCTTCCGATCTTGGAAAGGACGAAACACC |
| 10×-sgRNA i7-N724 | CAAGCAGAAGACGGCATACGAGATCGCTCAGTGTGACTGGAGTTCAGACG TGTGCTCTTCCGATCTTGGAAAGGACGAAACACC |
| 10×-sgRNA i7-N726 | CAAGCAGAAGACGGCATACGAGATGTCTTAGGGTGACTGGAGTTCAGACG TGTGCTCTTCCGATCTTGGAAAGGACGAAACACC |

5. Puromycin dihydrochloride.

6. Blasticidin hydrochloride.

7. Centrifuge with swinging bucket rotor and heating function.

8. BsmBI (10 U/μL).

9. Gibson assembly master mix (NEB).

10. NEBNext High-Fidelity 2× PCR Master Mix (NEB).

11. Endura ElectroCompetent cells (Lucigen).

12. MicroPulser Electroporator (Bio-Rad).

13. Gene Pulser Electroporation Cuvette: 0.1 cm gap (Bio-Rad).

14. Standard lysogeny broth (LB) medium with 100 μg/mL Ampicillin.

15. MinElute gel extraction kit (Qiagen).

16. Buffer EB (Qiagen).

17. ZymoPURE plasmid maxiprep kit (Zymo Research).

### 2.3 Virus Packaging, Titration, and Infection

1. OPTI-MEM reduced serum media (Thermo Fisher Scientific).

2. Linear polyethylenimine (PEI, Polysciences).

3. Hexadimethrine bromide (Polybrene, Sigma).

4. 24-well cell culture plates.

5. 0.45 μm filter and syringe.

6. Lenti-X lentivirus concentrator (Clontech).

7. CellTiter-Glo Luminescent Cell Viability Assay Kit (Promega).

8. 96-well plates with white wall and clear bottom.

9. Luminescence plate reader.

**2.4 Single-Cell RNA-Seq Library Construction**

1. Chromium Controller (10× genomics).
2. Chromium Single Cell 3′ Library & Gel Beads Kit (10× genomics).
3. Chromium Single Cell A Chip Kit (10× genomics).
4. Chromium i7 Multiplex Kit (10× genomics).
5. 0.4% Trypan Blue (Thermo Fisher Scientific).
6. 50% glycerol.
7. 8-channel multi-channel pipette.
8. 1 mL wide-orifice filtered tips (Rainin).
9. Molecular grade nuclease-free water (Sigma).
10. Molecular grade Bovine Serum Albumin solution, non-acetylated (Sigma).
11. SPRIselect Beads (Beckman Coulter).
12. Dynabeads MyOne Silane (Thermo Fisher Scientific).
13. 40 μM cell strainer.
14. Agilent TapeStation Automated Electrophoresis System: with D1000 and D5000 DNA ScreenTapes and reagents (Agilent).
15. Qubit Fluorometric Quantitation and dsDNA HS and broad-range kit (Thermo Fisher Scientific).
16. KAPA Library Quantification Kit for illumina libraries (KAPA Biosystems).

**2.5 Softwares for Data Processing**

1. BLAST (Basic Local Alignment Search Tool, NCBI).
2. Bedtools (http://bedtools.readthedocs.io/en/latest/).
3. Cell Ranger 2.1.1 (10× genomics).
4. The analysis pipeline (https://github.com/russellxie/Mosaic-seq2) (Fig. 2).

# 3 Methods

**3.1 Construction of sgRNA Library**

*3.1.1 Design and Synthesis of sgRNA Oligos*

To maximize the suppression efficiency and minimize the off-target of CRISPR system, we use the following steps to design our sgRNA library.

1. Select enhancers of interest based on epigenetic features (such as histone marks, eRNA transcription, P300 binding etc.) (Fig. 3). Overlap the enhancers of interests with the DNase-seq signals. The CRISPRi-target region is ±200-bp from the summit of the DNase-seq peak.

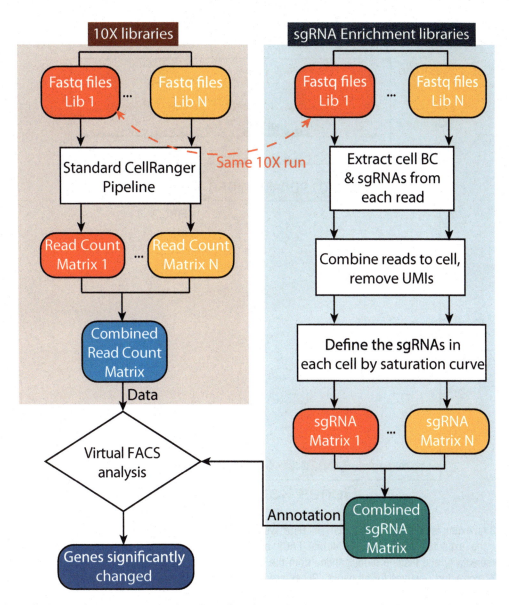

**Fig. 2** Overview of the analysis pipeline. Fastq files from 10× libraries and sgRNA enrichment libraries are processed separately, and then applied for differential gene expression analysis

2. Identify all the usable protospacer sequence on human genome based on the protospacer adjacent motif (PAM, "NGG" for Cas9). Nineteen base-pair protospacer is used to minimize the off-target [12].

3. Filter candidate sequences so that the closest secondary match on human genome is an edit distance of 2-bp away. To achieve this, compare each candidate sequence to the human genome by using BLAST (*see* **Note 2**).

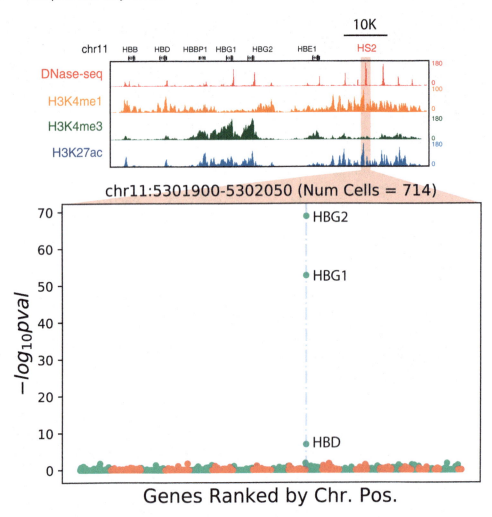

**Fig. 3** An example of Mosaic-seq2 analysis. Here, we use the beta-globin LCR in K562 cells as an example to show the expected results of the Virtual FACS analysis. The upper panel shows profiles of DNase I hypersensitivity and histone modifications from the ENCODE Project. The bottom panel is a Manhattan Plot of FDR-corrected $p$-values for all genes. Genes are ordered based on their relative position on the chromosomes. Two different colors indicate different chromosomes. Note that HBG2 and HBG1 are the primary target genes of the HS2 enhancer in this region, which have the most significant $p$-values in this unbiased test

4. Overlap the protospacer list in **step 3** with enhancer regions using the "bedtools intersect" function. Select multiple protospacer per enhancer regions.

5. Synthesize a single-stranded sgRNA oligo library based on the selected protospacer sequences, using the format provided in "sgRNA Oligo Pool" (Table 1).

1. PCR amplify the sgRNA library oligo to make it double stranded (Table 1, *see* **Note 3**).
   Reagents:

| Reagent | Vol. (µL) |
|---|---|
| NEBNext High-Fidelity 2× PCR master mix | 25 |
| Oligo Amp Fwd (10 µM) | 2 |
| Oligo Amp Rev (10 µM) | 2 |
| sgRNA oligo pool (10 µM) | 2 |
| $H_2O$ | 24 |

PCR Condition:

| Temp (°C) | Time | Cycles |
|---|---|---|
| 98 | 30 s | 1 |
| 98<br>65<br>72 | 10 s<br>30 s<br>30 s | 12 |
| 72 | 5 min | 1 |
| 4 | Hold | |

2. Run the PCR product on a 1% Agarose gel. Cut and purify the PCR product (~120 bp) by using a Qiagen MinElute kit. Elute the Oligo in 2 × 8 µL of EB buffer.

3. Combine all the reactions into one tube and measure DNA concentration with a Qubit fluorometer (Broad range DNA kit).

4. Digest 30 µg CROP-Guide-puro-bar plasmid with BsmBI for 10 h at 55 °C. To get the maximized cutting efficiency, 6 reactions are performed in parallel. Each reaction contains 5 µg of plasmid and 3 µL of enzyme.

5. Gel purify the linearized plasmid. Measure the concentration with a Qubit fluorometer (Broad range DNA kit).

6. Ligate the barcode oligo to the linearized plasmid by Gibson Assembly (GA). Use 2 µg PCR product and 3 µg digested plasmid in a 400 µL of GA reaction.

7. Incubate the mixture at 50 °C for 1 h.

8. Perform Qiagen MinElute cleanup and elute in 15 µL $H_2O$.

9. Measure the concentration with a Qubit fluorometer (Broad range DNA kit). Adjust the concentration to 1 µg/µL. If the concentration is lower than 1 µg/µL, perform SpeedVac to decrease the total volume.

*3.1.3 Electroporation*

1. Prechill the cuvette on ice.

2. Thaw and transfer 25 μL Endura Electrocompetent cells to the prechilled cuvette.

3. Add 1 μg GA product to the competent cells. The total volume of GA product should be less than 2 μL.

4. Place cuvette into *E. coli* Pulser Transformation Apparatus and electroporate at 1.8 kV (or follow the protocol if other elector-transformation pulser is used).

5. Quickly add 975 μL of the recovery medium into the cuvette and pipette up and down three times to resuspend the cells.

6. Transfer all the cells to a 1.5 mL tube and incubate at 37 °C for 1 h with rotation.

7. Serially dilute 10 μL of the transformation mixture in recovery medium 4 times, using a dilution factor of 10 each time.

8. Spread 10 μL of each dilution onto a LB-Ampicillin plate.

9. For the rest of the mixture, inoculate into a 500 mL flask containing 200 mL LB with Ampicillin.

10. Incubate the plates overnight at 30 °C, and the LB at 30 °C with shaking (*see* **Note 4**).

11. Collect the bacteria from the liquid culture and extract the plasmids by using the Zymo Maxiprep Kit.

12. Count the colony numbers on the plates to estimate the transformation efficiency (*see* **Note 5**).

*3.1.4 NGS Evaluation of Library Complexity*

1. Amplify the sgRNA fragment from the plasmid library, by using the following conditions (see Table 1 for the primer sequences):

Reagents:

| Reagent | Vol. (μL) |
| --- | --- |
| NEBNext High-Fidelity 2× PCR master mix | 25 |
| LKO1_5 primer (10 μM) | 1 |
| sgRNA_amp primer (10 μM) | 1 |
| Plasmid library | 100 ng |
| H₂O | Up to 50 μL |

PCR Condition:

| Temp (°C) | Time | Cycles |
|---|---|---|
| 98 | 30 s | 1 |
| 98<br>60<br>72 | 20 s<br>20 s<br>30 s | 15 |
| 72 | 3 min | 1 |
| 4 | Hold | |

2. Run the PCR product on 1% agarose gel. Excise the fragment at ~125 bp.

3. Perform Qiagen MinElute Gel cleanup, elute in 2 × 10 μL EB.

4. Spec DNA with a Qubit fluorometer (High Sensitivity DNA kit).

5. Use about 10 ng DNA for the second PCR step below to add the Illumina sequencing adapter. Pick different i7 indices if multiple libraries will be pooled for sequencing. Mix the staggered i5 primers to increase the complexity of the library (*see* **Note 6**, see Table 1 for the oligo sequences).

Reagents:

| Reagent | Vol. (μL) |
|---|---|
| NEBNext High-Fidelity 2× PCR master mix | 25 |
| sgRNA-Lib staggered mix i5 (10 μM) | 1 |
| sgRNA-Lib primer i7 (10 μM) | 1 |
| 10 ng DNA from **step 25** | 5 |
| H₂O | 18 |

PCR Condition:

| Temp (°C) | Time | Cycles |
|---|---|---|
| 98 | 30 s | 1 |
| 98<br>60<br>72 | 20 s<br>20 s<br>30 s | 15 |
| 72 | 3 min | 1 |
| 4 | Hold | |

6. Run samples on 1% agarose gel. Excise the fragment at ~350 bp.

7. Spec samples with a Qubit fluorometer (High Sensitivity DNA kit) and TapeStation D1000 ScreenTape.

8. Sequence the libraries on Illumina sequencer. Use paired-end sequencing. Use index 1 (P7) to demultiplex the libraries. The primer we used on the P5 end does not contain the actual index sequence. Read 1 will be a constant sequence (therefore is not required). Read 2 will consist of a universal primer (18 bp) followed by the sgRNA (20 bp).

**3.2 Lentivirus Packaging, Titration, and Infection**

*3.2.1 Virus Packaging*

1. One day before transfection, seed $8 \times 10^6$ 293T cells to 10 cm dish in order to get about 80% confluence in the next day. Each dish should have 10 mL complete medium.

2. One hour before transfection, change the medium to 10 mL fresh complete medium.

3. Transfect 293T cells with lentivirus package plasmids in the following ratio:

    (a) 4.5 μg pMD2.G, 7 μg psPAX2 and 9 μg CROP-seq sgRNA library per dish.

    (b) For transfection, add all plasmids to 1 mL OPTI-MEM, and then add 61 μL PEI (1 mg/mL). Mix up gently.

    (c) Incubate at room temperature for 15 min.

    (d) And add the mixture into cells drop-wisely. Mixing by shaking the dish gently.

4. Place the plate back to an incubator.

5. Twelve hours after transfection, change the medium to 10 mL complete medium.

6. About 72 h after transfection, collect the supernatant from the plate.

7. Filter the supernatant by using 0.45 μm filter to remove cells and debris.

8. For concentration and purification of the virus, combine the medium from different dishes, add 1/3 volume of Lenti-X Concentrator reagent. Incubate at 4 °C overnight.

9. Centrifuge the sample at $1500 \times g$ for 45 min at 4 °C. After centrifuging, an off-white pellet should be visible.

10. Carefully remove all the supernatant, resuspend the pellet with 1/10 or 1/100 of the original volume.

11. Aliquote the virus into 1.5 mL tubes and freeze the virus at −80 °C.

*3.2.2  Virus Titration*

1. Seed $2 \times 10^5$ K562 cells to a 96-well plate (*see* **Note 7**).

2. Thaw one virus aliquot. In a separate 96-well plate, serial dilute 20 μL virus into 180 μL complete medium with 8 μg/μL polybrene five times with dilution factor of 10 ($10^1 \times – 10^5 \times$), each dilution has three replicates.

3. Briefly centrifuge the 96-well plate with cells, gently aspirate the supernatant while keeping the K562 cells at the bottom.

4. Transfer the diluted virus to the plate containing cells. Also keep several wells uninfected as the negative control.

5. Centrifuge at $1000 \times g$, 36 °C for 60 min.

6. Return the plate to the incubator.

7. The next day, change the medium in each well to the complete medium with 1 μg/mL puromycin. Incubate for another 3–4 days.

8. Monitor the cell growth in the next 3–4 days until all the cells in the uninfected wells are dead.

9. Dilute the CellTiter-Glo reagent with equal volume of PBS.

10. Aspire the medium in each well while keeping most of the cells in the well. Add 50 μL of the diluted CellTiter-Glo in each well. Incubate at room temperature for 10 min with mild shaking.

11. Transfer the mixture to a new 96-well plate with white wall and clear bottom for plate reading. Read the luminescence intensity of the whole plate. Instrument settings depend on the manufacturer. An integration time of 0.25–1 s per well is recommended.

12. Calculate the virus titer by using the dilution that has cell survival rate between 0% and 20%. Based on the Poisson Distribution, with the survival rate at this range, >90% of the infected cells will be infected by only one virus. Therefore, the survival cell rate can be used to estimate the virus titer.

*3.2.3  Virus Infection*

1. Plate $2 \times 10^5$ K562 cells into one well of a 24-well plate, which contains 500 μL of complete medium and 8 μg/mL polybrene (*see* **Note 8**).

2. Add desired amount of virus into each well (*see* **Note 9**).

3. Centrifuge the plate at 36 °C, $1000 \times g$ for 1 h.

4. Return the plate to the incubator.

5. Change the medium to a fresh complete medium with antibiotics. For K562 cells, we use 20 μg/mL for blasticidin and 1 μg/mL for puromycin. Combine and transfer the cells to a 75 cm$^2$ flask (*see* **Note 10**).

6. Keep the cells growing for the next 5–7 days. Split the cells once it reaches confluency. The cells should grow robustly in the presence of antibiotics.

**3.3 Construction of Single-Cell RNA-Seq Libraries**

The library construction follows the standard protocol of 10× Genomics Single Cell 3′ Library kit with several modifications. For the following steps, we will primarily focus on the modified steps. The standard steps are labeled as "*Std.*" For more details, please check the 10× genomics website (*see* **Note 11**). https://support.10xgenomics.com/single-cell-gene-expression/library-prep/.

1. Infect K562 cells with dCas9-KRAB virus, culture the cells for 2 weeks under the selection of blasticidin to generate a stable cell line expressing dCas9-KRAB. The cells can be frozen for future experiments.

2. Infect K562-dCas9-KRAB cells with the virus containing sgRNA library and NC sgRNAs (sgNC), respectively. Culture the cells for 5–7 days under selection of puromycin.

3. Transfer the infected K562-dCas9-KRAB cells to 50 mL conical tubes and centrifuge at $500 \times g$ for 4 min. Aspirate the medium and wash once with PBS. Process the sgNC cells and library infected cells separately (*see* **Note 12**).

4. Resuspend the cells in a small volume (*see* **Note 13**) of PBS + 0.01% BSA (PBS-BSA, non-acetylated). Gently resuspend the cells by using a wide bore 1 mL tip.

5. Pass the cells through 40 µL strainer to get rid of debris (*see* **Note 14**).

6. Count the cell concentration by using Trypan Blue. Dilute the cell appropriately before counting since they may be over 1000 cells/µL. Take both live and dead cells into account. The cells must be in single-cell suspension and the survival rate should be over 90% (this is very important if adherent cells are used).

7. (*Std*) For generating a library of 10,000 cells, dilute the cells to 1000 cells/µL (*see* **Note 15**) by using PBS-BSA.

8. Mix the 5% NC cells with library infected cells (1:20). Keep the cells on ice while processing other reagents.

9. (*Std*) Load the cells to the Single Cell A Chip, as well as the partition oil and beads.

10. (*Std*) Run the Chromium Controller following the instruction. Transfer the GEMs to new strip PCR tubes.

11. (*Std*) Perform reverse transcription reactions for GEMs.

12. (*Std*) Clean up the cDNA by using the Silane DynaBeads.

13. (*Std*) Perform the cDNA amplification PCR.

14. (*Std*) Perform SPRI Beads cleanup for the cDNA product. Elute in 40.5 µL of Buffer EB.

15. (*Std*) Run 1 µL of sample on Agilent TapeStation D5000 ScreenTape to determine the cDNA size distribution. Also use 1 µL of sample in the Qubit Fluorometer (High Sensitivity DNA kit) to accurately quantify concentration.

16. Save 15 ng of cDNA for the sgRNA enrichment PCR (about 5 µL). Add 5 µL of Buffer EB to the original cDNA sample and continue the standard 10× library prep.

17. (*Std*) Finish the standard 10× library preparation.

18. Use the 15 ng cDNA for the enrichment PCR of sgRNA. Choose primers with different i7 indices for different libraries.

Reagents:

| Reagent | Vol. (µL) |
| --- | --- |
| NEBNext High-Fidelity 2× PCR master mix | 25 |
| SI-PCR Primer (10 µM) | 1 |
| 10×-sgRNA primer i7 (10 µM) | 1 |
| 15 ng DNA from **step 15** | |
| $H_2O$ | to 50 |

PCR Condition:

| Temp (°C) | Time | Cycles |
| --- | --- | --- |
| 98 | 45 s | 1 |
| 98 | 20 s | 15 |
| 54 | 30 s | |
| 72 | 20 s | |
| 72 | 1 min | 1 |
| 4 | Hold | |

19. Perform 0.8× SPRI Beads clean up by adding 40 µL beads to the PCR mixture from last step. Elute in 20 µL Buffer EB.

20. Perform Qubit quantitation (High Sensitivity DNA kit) and TapeStation on both the original 10× libraries and the enrichment libraries.

21. Before loading on the sequencer, perform qPCR on all the libraries by using the KAPA library quantification kit. Calculate the nanomolar concentration of each library.

22. Mix the libraries based on the desired number of reads need to be sequenced. Consulting the sequencing core for the final concentration of library required (*see* **Note 16**).

**3.4  Data Analysis**

*3.4.1  Mapping of 10×*
*Libraries*

We use the Cell Ranger software from 10× genomics for preprocessing of the single-cell sequencing data. For more information,
please check the 10× website (https://support.10xgenomics.com/
single-cell-gene-expression/software/pipelines/latest/what-is-cell-
ranger).

1. Use "cellranger mkfastq" command to create FASTQ files
   from the raw imaging files with the flag "--ignore-dual-index."

2. Use "cellranger count" command to map and preprocess the
   data, with the flag "--expect-cells = 10,000."

3. To combine multiple 10× libraries from different sequencing
   runs, use "cellranger aggr" command with the flag "--
   normalize = mapped."

4. The combined HDF5 file named "filtered_gene_bc_matri-
   ces_h5.h5," which contains all the read count matrices, gene
   names, and cell barcodes, will be used for the following
   analysis.

*3.4.2  Processing*
*of sgRNA Enrichment*
*Libraries*

We have established a Github repository detailing a computational
pipeline to process sgRNA enrichment libraries (https://github.
com/russellxie/Mosaic-seq2). Here, we describe an overview of
this pipeline.

1. Read 1 represents the cell barcode and UMI. Read 2 represents
   the sgRNA transcript sequence.

2. Using the lists of known sgRNAs and cell barcodes (output
   from the 10× pipeline), filter the reads for those with known
   barcodes and sgRNAs. The possible match position(s) of each
   known sgRNA/barcode on the read was identified by using the
   "regex" package of python with edit distance of 1.

3. Count the number of UMIs for each sgRNA in each cell using
   the "directional" method described in UMI-tools [13].

4. Determine the number of sgRNAs in each cell using a satura-
   tion curve based method described in [14]. Briefly, for a given
   sgRNA, define the set C as the UMI count in all the cells with
   this sgRNA. Then identify the inflection point $N$ of this sgRNA
   by the cardinality of set $A$, given:

$$N = n(A), A = \left\{ C_i \middle| \left( \frac{C_i}{U} > \frac{1}{\nu} \right) \right\}$$

where $C_i$ is a given element in set $C$, $U$ denotes the total number of
UMIs for this sgRNA, and $\nu$ represents the total number of cells
containing this sgRNA. Then $N$ is a cutoff of the UMI count for
this sgRNA in every single cell. Adjust the UMI count to 0 if the
observed count is equal to or smaller than $N$. The sgRNAs with the
adjusted UMI count which is greater than 0 will be considered a
real sgRNA in this cell.

*3.4.3 Virtual FACS Analysis*

1. We only consider the genes expressed at least in 5% of the cells.

2. Define M as the total number of cells in the population.

3. Define $E_g$ as the median expression (in units of cpm) of a given gene $g$.

4. Define $N$ as the number of cells with expression less than $E_g$. Note that because of the zero-inflated nature of scRNA-Seq data, $N$ is often not $M/2$.

5. Define $K_s$ as the number of cells expressing sgRNA $s$ (*see* **Note 17**).

6. Define $X_{g,s}$ as the number of cells expressing $s$ that have expression of $g$ less than $E_g$.

7. For each gene $g$ and sgRNA $s$, calculate a p-value using a hypergeometric test with parameters $M$, $N$, $X_{g,s}$ and $K_s$.

8. Adjust the *p*-values of each sgRNA/enhancer region based on Benjamini-Hochberg procedure (FDR) (Fig. 3).

# 4    Notes

1. We order individual oligos from Integrated DNA Technologies (IDT). All the Nextera indexing primers were purified by using HPLC. We order sgRNA oligo pools from CustomArray.

2. To BLAST all the possible protospacer sequence on human genome takes very long time even on a HPC system. However, once this step is completed, a full list of all usable CRISPR protospacer sequences will be generated and can be used in other studies.

3. Estimate the number of PCR reactions based on the total oligos required in the Gibson Assembly step. The yield of DNA is about 100–150 ng/reaction if using the conditions described here.

4. Growing *E. coli* at 37 °C may increase recombination in the lentiviral construct. Alternatively, bacteria can be grown on large bioassay dishes (From Thermo Fisher Scientific).

5. The efficiency of Endura competent cells can reach $1 \times 10^{10}$ cfu/μg if more than 1 μg of DNA is used per electroporation. Colony numbers should be at least $30\times$ of the number of sgRNAs in the library to maintain the complexity.

6. We only list primers with N720-726 in Table 1. Other Nextera i7 indices are also compatible.

7. Virus titration should be performed on the same cell line as Mosaic-seq experiments to get the best estimation.

8. The numbers of wells used for infection depends on the complexity of the sgRNA library. We recommend infecting at least

100× cells as the number of sgRNAs in the library at high MOI. For low MOI infection, this number needs to be increased.

9. To attain ~1 sgRNA per cell, we infect at MOI = 0.2. In our hands, an MOI of 5 can give an average of 2–3 sgRNAs detected per cell in single-cell RNA-seq.

10. The concentration of antibiotics needs to be defined for each cell type. Perform a serial dilution of antibiotics and pick the lowest concentration that could kill most of the cells.

11. Other single-cell RNA-seq methods such as Drop-seq, InDrops are also compatible as long as the assay captures the 3′-end of the mRNAs.

12. The actual cells required for this step depend on the scale of the experiment. Each 10× lane could generate a library containing about 10,000 cells, with about 7% doublet rate. Currently we sequence ~50 cells per sgRNA.

13. Resuspend in a small volume of PBS-BSA so that the final concentration of cells is 800–1000 cells/μL.

14. If the volume of cells is smaller than 500 μL, use the Flowmi Tip Strainers (SP Scienceware).

15. Follow the 10× protocol for dilution. If the cell concentration is lower than 600 cells/μL, a single run cannot reach 10,000 cells.

16. We normally sequence 10–20 million reads for the enrichment libraries and 200 million reads for each 10× libraries (with 10,000 cells). These numbers may be adjusted based on the numbers of cells and number of total sgRNAs in the library. The enrichment library can be sequenced by using the same configuration with the standard 10× libraries.

17. This analysis can be performed per enhancer regions or per sgRNA depending on how many cells are sequences.

## Acknowledgments

This work is supported by the Cancer Prevention Research Institute of Texas (CPRIT) (RR140023, G.C.H.), NIH (DP2GM128203, G.C.H.), the Department of Defense (PR172060, G.C.H.), the Welch Foundation (I-1926-20170325, G.C.H.), and the Green Center for Reproductive Biology. S.X. is an American Heart Association fellow (16POST29910007).

## References

1. ENCODE Project Consortium (2012) An integrated encyclopedia of DNA elements in the human genome. Nature 489:57–74

2. Kundaje A, Meuleman W, Roadmap Epigenomics Consortium et al (2015) Integrative analysis of 111 reference human epigenomes. Nature 518:317–330

3. Long HK, Prescott SL, Wysocka J (2016) Ever-changing landscapes: transcriptional enhancers in development and evolution. Cell 167:1170–1187

4. Xie S, Duan J, Li B et al (2017) Multiplexed engineering and analysis of combinatorial enhancer activity in single cells. Mol Cell 66:285–299.e5

5. Gilbert LA, Larson MH, Morsut L et al (2013) CRISPR-mediated modular RNA-guided regulation of transcription in eukaryotes. Cell 154:442–451

6. Thakore PI, D'Ippolito AM, Song L et al (2015) Highly specific epigenome editing by CRISPR-Cas9 repressors for silencing of distal regulatory elements. Nat Methods 12:1143–1149

7. Dixit A, Parnas O, Li B et al (2016) Perturb-seq: dissecting molecular circuits with scalable single-cell RNA profiling of pooled genetic screens. Cell 167:1853–1866.e17

8. Jaitin DA, Weiner A, Yofe I et al (2016) Dissecting immune circuits by linking CRISPR-pooled screens with single-cell RNA-seq. Cell 167:1883–1896.e15

9. Datlinger P, Rendeiro AF, Schmidl C et al (2017) Pooled CRISPR screening with single-cell transcriptome readout. Nat Methods 14:297–301

10. Hill AJ, McFaline-Figueroa JL, Starita LM et al (2018) On the design of CRISPR-based single-cell molecular screens. Nat Methods 15:271–274

11. Xie S, Cooley A, Armendariz D et al (2018) Frequent sgRNA-barcode recombination in single-cell perturbation assays. PLoS One 13: e0198635

12. Xu H, Xiao T, Chen C-H et al (2015) Sequence determinants of improved CRISPR sgRNA design. Genome Res 25:1147–1157

13. Smith T, Heger A, Sudbery I (2017) UMI-tools: modeling sequencing errors in unique molecular identifiers to improve quantification accuracy. Genome Res 27:491–499

14. Macosko EZ, Basu A, Satija R et al (2015) Highly parallel genome-wide expression profiling of individual cells using nanoliter droplets. Cell 161:1202–1214

# Chapter 15

# Antigen Receptor Sequence Reconstruction and Clonality Inference from scRNA-Seq Data

## Ida Lindeman and Michael J. T. Stubbington

## Abstract

In this chapter, we describe TraCeR and BraCeR, our computational tools for reconstruction of paired full-length antigen receptor sequences and clonality inference from single-cell RNA-seq (scRNA-seq) data. In brief, TraCeR reconstructs T-cell receptor (TCR) sequences from scRNA-seq data by extracting sequencing reads derived from TCRs by aligning the reads from each cell against synthetic TCR sequences. TCR-derived reads are then assembled into full-length recombined TCR sequences. BraCeR builds on the TraCeR pipeline and accounts for somatic hypermutations (SHM) and isotype switching. Here we discuss experimental design, use of the tools, and interpretation of the results.

**Key words** TCR, BCR, Immunoglobulin, Single cell, RNA-seq, scRNA-seq, Antigen receptor reconstruction, Tracer, Bracer

## 1  Introduction

T cells and B cells recognize antigens in a highly specific manner through their cell-surface T-cell receptor (TCR) or B-cell receptor (BCR). These receptors are extremely diverse heterodimers comprising a TCRα- and a TCRβ-chain (αβ T cells), a TCRγ- and a TCRδ-chain (γδ T cells) or a heavy (IgH) and a light (Igκ or Igλ) chain (B cells) encoded by genes generated through V(D)J recombination during the development of the cell in the thymus or bone marrow. High-throughput antigen receptor sequencing (Rep-seq) of a single type of chain in bulk populations has been a common strategy for characterization of BCR- and TCR-repertoires [1–3], but lacks information about chain pairing (reviewed in [4, 5]). Single-cell Rep-seq is useful to decipher paired antigen receptor repertoires, but provides very limited additional information about the cells. While Rep-seq in combination with phenotyping primers can give information on the expression of a selected panel of genes in addition to the paired antigen receptor [6], sequencing of the

Guo-Cheng Yuan (ed.), *Computational Methods for Single-Cell Data Analysis*, Methods in Molecular Biology, vol. 1935, https://doi.org/10.1007/978-1-4939-9057-3_15, © Springer Science+Business Media, LLC, part of Springer Nature 2019

entire transcriptome is a much more informative and unbiased approach (reviewed in [7]).

During the last few years, single-cell RNA-sequencing (scRNA-seq) has been an extremely valuable approach for identifying and characterizing heterogeneity in cell subsets and cells in various differential states both in health and in disease (reviewed in [8]). The development of computational tools allowing researchers to reconstruct TCR- and BCR-sequences directly from scRNA-seq data thus provides a unique opportunity to gain insight into T- and B-cell immunity, cell fate, lineage evolution, and antigen-specific responses by linking antigen receptor usage to the full transcriptomic identity of individual T- or B-cells.

Here we describe TraCeR [9] and BraCeR [10], our computational tools for reconstruction of paired full-length antigen receptor sequences and clonality inference from scRNA-seq data. During the last 2 years several other tools for TCR- and/or BCR-reconstruction from scRNA-seq data have emerged [11–15], illustrating that there is a high level of interest in approaches such as these.

TraCeR reconstructs TCR sequences from scRNA-seq data by extracting sequencing reads derived from TCRs by aligning the reads from each cell against synthetic TCR sequences representing all possible combinations of V- and J-segments. TCR-derived reads are then assembled into full-length TCR sequences. BraCeR builds on the TraCeR pipeline, accounting for somatic hypermutations (SHM) and isotype switching.

We have previously demonstrated an application for TraCeR by investigating CD4+ T-cell clonotypes in the spleen of mice as a response to a *Salmonella* infection [9] where members of each expanded T-cell clone were found across various proliferation and differentiation states. More recently, the use of TraCeR in combination with pseudotime and branching inference revealed that the progeny of a single naïve murine CD4+ T cell can be found in both $T_H1$ and $T_{FH}$ compartments during an immune response to malaria [16]. Patil et al. demonstrated clonal sharing between a population of CD4+ cytotoxic T lymphocyte (CD4-CTL) precursors and effector memory CD4-CTLs [17]. Furthermore, TraCeR has been used to map regulatory T-cell (Treg) clones and memory T-cell clones across lymphoid and non-lymphoid tissues in a study focusing on identifying trajectories of tissue adaptation [18].

## 2 Materials

In this chapter we use the notions "[TB]raCeR" ("TraCeR" or "BraCeR") and "[tb]racer" ("tracer" or "bracer") in order to avoid unnecessary repetitions when describing both tools.

***2.1    Sequencing Data***    The following aspects should be taken into consideration when choosing a library preparation protocol and sequencing platform for the generation of data as input to [TB]raCeR.

1. Choose a library preparation protocol that generates sequencing reads from the full length of mRNA transcripts (*see* **Note 1**).

2. Paired-end (PE) reads provide the maximum reconstruction rate and accuracy compared with single-end (SE) reads.

3. Sequence your library with a minimum read length of 50 bases (*see* **Note 2**).

4. The read depth required to reconstruct TCRs or BCRs from a single cell depends on the cell type and activation state (*see* **Note 3**).

5. Make sure the reads are demultiplexed according to the cell of origin after sequencing.

6. Perform basic quality control of the raw reads (*see* **Note 4**).

7. [TB]raCeR accepts FASTQ files (fastq or fastq.gz) as input. BraCeR also accepts assembled BCR sequences in FASTA format for clonality inference (*see* Subheading 3.2.4).

***2.2    Prerequisites***

*2.2.1    External Tool Requirements for [TB] raCeR*

1. Python (≥2.7.0) (*see* **Note 5**).

2. Bowtie 2 [19].

3. Trinity [20] (*see* **Note 6**).

4. IgBLAST [21] (*see* **Note 7**).

5. Kallisto [22] or Salmon [23] (*see* **Note 8**).

6. Graphviz (*see* **Note 9**).

*2.2.2    Additional Prerequisites Specific to BraCeR (See Note 10)*

1. Python (≥3.4.0).

2. BLAST [24].

3. Trim Galore!

4. PHYLIP dnapars.

5. R (≥3.1.2) and R packages for lineage reconstruction: ggplot2, Rscript, Alakazam.

***2.3    Installing [TB] raCeR***

*2.3.1    Installing [TB] raCeR from GitHub*

1. Download or clone the GitHub repository (https://github.com/Teichlab/tracer or https://github.com/Teichlab/bracer) with *git clone*.

2. Install all required prerequisites.

3. Set up Python dependencies (*see* **Note 11**).

4. Install tracer/bracer module with *python setup.py install* (*see* **Note 12**).

5. Edit configuration file (*see* **Note 13**).

*2.3.2 Running [TB]raCeR as a Standalone Docker Image*

Alternatively, [TB]raCeR can be run as a standalone Docker image on DockerHub, with all of the dependencies installed and configured appropriately (*see* **Note 14**).

1. Pull the Docker container from DockerHub with *docker pull teichlab/[tb]racer.*

2. Increase the memory limit for Docker to 6–8 GB (*see* **Note 15**).

3. Run the following command, followed by any appropriate arguments, from your input data directory: *docker run -it --rm -v $PWD:/scratch -w /scratch teichlab/[tb]racer.*

*2.4 Testing [TB] raCeR*

1. Run [*tb*]*racer test* with optional arguments (*see* **Note 16**).

2. Compare the output in *test_data/results/filtered_[TB]CR_summary* with the expected results in *test_data/expected_summary* (*see* **Note 17**).

# 3   Methods

*3.1 [TB]raCeR Pipeline*

The [TB]raCeR pipelines consist of two main steps (Fig. 1):

1. Reconstruction of TCR/BCR sequences from each cell (*assemble* command).

2. Creation of clonal networks (*summarise* command).

*3.2 Reconstruction of TCR/BCR Sequences with Assemble*

*3.2.1 Overview of Pipeline*

The *assemble* stage performs the steps:

1. Trimming of raw reads to remove adapter sequences and low-quality sequences (BraCeR only, *see* **Note 18**).

2. Extract TCR/BCR-derived reads by alignment to a combinatorial recombinome using Bowtie 2 [19] (*see* **Note 19**).

3. Perform a second round of alignment for IgH if the reads are 50 bases or shorter to extract reads mapping mainly or solely to the CDR3 (*see* **Note 20** and Fig. 2).

4. Assemble TCR/BCR-derived reads into contigs using Trinity.

5. Detect isotype (BraCeR only, *see* **Note 21**).

6. Determine productivity and gene usage of reconstructed sequences.

7. Collapse highly similar sequences.

8. Quantify the expression of each reconstructed sequence and filter based on expression.

*3.2.2 Preparing Input for [TB]raCeR*

[TB]raCeR takes as input FASTQ files containing sequencing reads generated from a *single* cell. Thus, data must be demultiplexed such that reads from each cell are identified and written to separate files.

## Step 1: Reconstruction of antigen receptor sequences

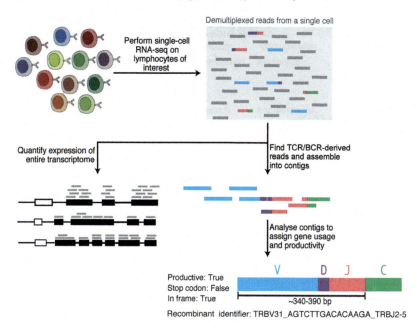

## Step 2: Clonality analysis

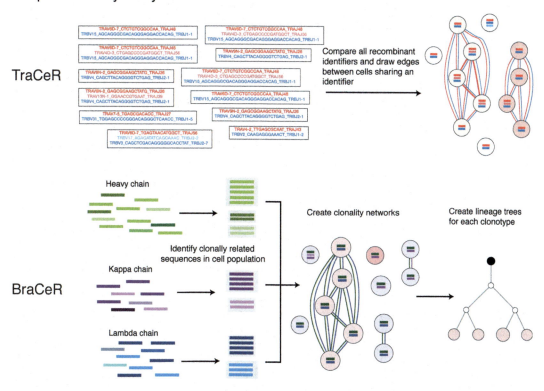

**Fig. 1** Overview of the [TB]raCeR pipelines. Adapted from Stubbington et al. (ref. 9)

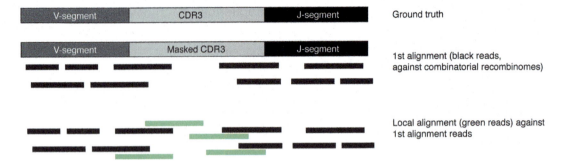

**Fig. 2** Illustration of the two alignment steps for IgH when reads are 50 bases or shorter

Data can be paired-end (PE) or single-end (SE) with PE providing higher sensitivity sequence reconstruction.

Although it is beyond the scope of this chapter to provide instructions for quality control of scRNA-seq data, we recommend that users exclude apparently poor-quality cells from any downstream analyses. This can be done using automated methods [25] or by manual inspection of metrics such as total number of mapped reads, percentage of reads mapping to the mitochondrial genome, rate of mapping to the transcriptome, and number of genes detected.

*3.2.3 Running [TB]raCeR in Assemble Mode with Default Settings*

Run [TB]raCeR *assemble* with the main arguments described below. Note that the two tools have slightly different usage (*see* below), but many of the arguments are the same.

> *tracer assemble [options] <file_1> [<file_2>] <cell_name> <output_directory>*
>
> *bracer assemble [options] <cell_name> <output_directory> [<file_1>] [<file_2>]*

1. *<file_1>* is the FASTQ file providing #1 mates from PE sequencing or all of the reads from SE sequencing. May be left blank if running BraCeR with *--assembled_file*.

2. *<file_2>* is the FASTQ file providing #2 mates for PE reads.

3. *<cell_name>* is a name that will be used for references to the cell.

4. *<output_directory>* is the directory for output, and should be identical for cells to be summarized together.

*3.2.4 Running [TB]raCeR in Assemble Mode with Optional Arguments*

The following optional arguments can be passed to either TraCeR or BraCeR:

1. *-s/--species*. Species from which the cells were derived. The default is mouse (*Mmus*) for TraCeR and human (*Hsap*) for BraCeR.

2. *--loci*: Loci to reconstruct. Default is "A B" for TraCeR and "H K L" for BraCeR. Include or replace with "G D" to attempt to reconstruct TCRγ and TCRδ.

3. *--single_end*: Set this flag if your data are SE reads.

4. *--fragment_length*: The estimated average fragment length of the sequencing library. Required for SE data.

5. *--fragment_sd*: The estimated standard deviation of the average fragment. Required for SE data.

6. *-p/--ncores*: The number of processor cores to use. Default = 1.

7. *--resource_dir*: Path to directory containing resources required for alignment. Use if you wish to use other resources contained somewhere other than the default resources directory.

8. *-c/--config_file*: Path to the configuration file (*see* **Note 13**). Default = ~/.[tb]racerrc.

9. *-r/--resume_with_existing_files*: If this flag is set, [TB]raCeR will look for existing output files and skip already completed steps.

10. *--max_junc_len*: The maximum allowed length of junction string in a recombinant identifier (*see* **Note 22**).

The optional arguments below are specific to TraCeR:

1. *-m/--seq_method*: Method for generation of sequences for output and productivity assessment. Options are *-m imgt* (default) and *-m assembly* (*see* **Note 23**).

2. *-q/--quant_method*: Method used for expression quantification (*kallisto* or *salmon*).

3. *--small_index*: Set this flag for faster expression quantification if you have prebuilt a transcriptome index (*see* **Note 24**).

4. *--invariant_sequences:* Path to file specifying invariant sequences for particular unconventional T-cell types. For an example, see the default file at *resources/Mmus/invariant_cells.json*.

The optional arguments below are specific to BraCeR:

1. *--assembled_file*: Path to a FASTA file with pre-assembled sequences for a cell. Make BraCeR skip the alignment and assembly steps (*see* **Note 25**).

2. *--no_trimming:* Do not remove adapter sequences and low-quality reads.

3. *--keep_trimmed_reads:* Keep the output files from the trimming step.

*3.2.5 The Sequence Identifier Format*

In order to facilitate comparison of sequences between cells, the IgBLAST results for each TCR/BCR sequence within a cell are represented by a sequence identifier string (e.g.,

TRBV31_AGTCTTGACACAAGA_TRBJ2–5). This sequence identifier format consists of:

1. Most likely V-gene name.

2. Junctional or CDR3 nucleotide sequence (*see* **Note 26**).

3. Most likely J-gene name.

In cases where the V- or J-gene assignments are uncertain, [TB] raCeR uses a list of all possible sequence identifiers for comparisons. The sequence identifiers are not used directly by BraCeR for clonality inference, but serve as additional information in the clonotype networks.

*3.2.6   Output of* Assemble

The output of the *assemble* step is an *<output_directory>/< cell_name>* directory for each cell, containing all or most of the following subdirectories:

1. *trimmed_reads*: Contains reads trimmed by Trim Galore! if *bracer assemble* is run with *--keep_trimmed_reads*.

2. *aligned_reads*: Contains Bowtie 2 output with TCR/BCR-derived reads.

3. *Trinity_output*: Contains FASTA files with assembled contigs for each locus, as well as two log files.

4. *IgBLAST_output*: Contains IgBLAST output for the contigs from each locus.

5. *BLAST_output*: Contains BLAST output for the contigs from each locus.

6. *unfiltered_[TB]CR_seqs*: Contains files detailing all reconstructed TCR/BCR sequences (*see* **Note 27**).

7. *expression_quantification*: Contains output of Kallisto/Salmon for the entire transcriptome and the TCRs/BCRs (*see* **Note 28**).

8. *filtered_[TB]CR_seqs:* Contains the two most highly expressed recombinants for each locus.

## 3.3   Creation of Clonotype Networks and Lineage Trees with Summarise

### 3.3.1   Defining Clonally Related Recombinants

Both productive and nonproductively rearranged TCR sequences are considered clonally related if they share a sequence identifier, meaning that they share assignment of V- and J gene and the junctional sequence. Clonally related productively rearranged BCR sequences are identified for each locus with the Change-O toolkit [26] based on the following criteria (Fig. 3):

1. Common V- and J-gene in the sets of potential V- and J-genes between the sequences.

2. Equal CDR3 length.

3. CDR3 nucleotide distance normalized by length <0.2 (*see* **Note 29**).

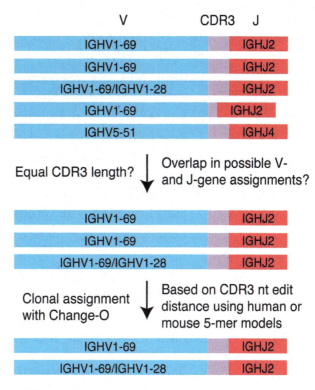

**Fig. 3** Identification of clonally related productive BCR sequences for each locus

*3.3.2 Generation of Clonal Networks*

We use custom scripts to assess the clonal groups and generate network graphs as follows.

1. Each single cell is represented by a node in the graph.

2. Reconstructed sequences are represented within nodes by horizontal lines colored according to locus and productivity or by the sequence identifier.

3. Edges between the nodes represent clonally related TCR/BCR sequences, and are color coded according to locus. Edges between B cells are only drawn if they share a clonally related productive IgH and a clonally related productive Igκ or Igλ (*see* **Note 30**).

4. Edge thickness is proportional to the number of shared sequences for a locus.

5. Nonproductively rearranged BCR sequences are determined to be shared within a clone group and included as edges in the graph if they have overlapping V- and J-gene assignments. If the cells only share a nonproductive chain for a specific locus (Igκ or Igλ), this is shown with a dotted instead of a solid line in the clonal network.

**3.3.3  Construction of Immunoglobulin Lineage Trees**

BraCeR offers a complete pipeline based on both heavy and light chains for construction of lineage trees through Change-O, Alakazam [27], and PHYLIP [28], consisting of the following:

1. Build IgBLAST reference databases using IMGT-gapped sequences (*see* **Note 31**).

2. Run IgBLAST on all sequences belonging to a clone group.

3. Parse IgBLAST output and create Change-O database.

4. Add clone number, isotype and cell name to the Change-O database for each sequence.

5. Reconstruct the germline sequences (with masked junction) in each clone group with Change-O *CreateGermlines*.

6. Concatenate productive heavy and light chain shared in each clone group (*see* **Note 32**).

7. Run the appropriate Alakazam commands through our *lineage. R* script (*see* **Note 33**).

**3.3.4  Running [TB]raCeR in Summarise Mode**

Run the [TB]raCeR *summarise* command with options as described below. *<input_dir>* is the directory containing subdirectories of each cell you want to summarise (*see* **Note 34**).

*[tb]racer summarise [options] <input_dir>*

The following optional arguments can be passed to either TraCer or BraCeR:

1. *-c/--config_file*: Path to the configuration file (*see* **Note 13**). Default = *~/.[tb]racerrc*.

2. *-u/--use_unfiltered*: Set this option to run *summarise* with all reconstructed recombinants without filtering cases where more than two sequences are detected for a particular locus.

3. *--resource_dir*: Path to directory containing resources required for alignment. Use if you wish to use other resources contained somewhere other than the default resources directory.

4. *-s/--species*: Species of origin. Default = *Mmus* (mouse) for TraCeR and *Hsap* (human) for BraCeR (*see* **Note 35**).

5. *--loci*: Space-separated list of loci to summarize (*see* **Note 36**).

6. *-g/--graph_format*: Output format of clone networks (*see* **Note 37**).

7. *--no_networks*: Do not draw clonotype network graphs (*see* **Note 38**).

The following optional arguments are specific to TraCeR:

1. *--receptor_name*: Specify if other than "TCR" when using the *Build* module.

2. *-i/--keep_invariant*: Set this option to keep invariant cells (*see* **Note 39**).

The following optional arguments are specific to BraCeR:

1. *--IGH_networks*: Base clonality solely on IgH, allowing clone groups with different or no light chain.

2. *--dist*: Distance value (float) for clonal inference. Default = 0.2 (*see* **Note 40**).

3. *--include_multiplets*: Set if you do not wish to exclude potential cell multiplets from downstream analyses.

4. *--infer_lineage*: Attempt lineage tree construction for clone groups.

*3.3.5 Output of the TraCeR Summarise Step*

The output of the TraCeR summarise step is written to *filtered_TCR<loci>_summary* or *unfiltered_TCR<loci>_summary*. The following output files are generated:

1. *TCR_summary.txt*: TCR reconstruction summary statistics file.

2. *recombinants.txt:* File listing the identifier, lengths, and productivity of each reconstructed TCR for each cell.

3. *reconstructed_lengths_TCR[A|B].[pdf|txt]*: Distribution plots and underlying data displaying reconstructed VDJ region lengths for each locus.

4. *clonotype_sizes.[pdf|txt]:* Distribution of clonotype sizes shown as bar plots and underlying data.

5. *clonotype_network_[with|without]_identifiers.<graph_format>*: Clonotype networks in graphical format with recombinant identifiers or with lines representing the presence of recombinants for a locus in a cell.

6. *clonotype_network_[with|without]_identifiers.dot*: Clonotype networks described in the Graphviz DOT language.

*3.3.6 Output of the BraCeR Summarise Step*

The following output files and subdirectories may be generated (depending on options):

1. *BCR_summary.txt:* BCR reconstruction summary statistics file.

2. *changeodb.tab:* Database file describing all reconstructed sequences in single cells. Recombinants in suspected multiplets are included if run with *--include_multiplets*.

3. *filtered_multiplets_changeodb.tab:* Database file with reconstructed recombinants from suspected multiplets unless run with *--include_multiplets*.

4. *IMGT_gapped.tab:* Database file for all reconstructed sequences based on IMGT-gapped reference sequences.

5. *reconstructed_lengths_BCR[H|K|L].[pdf|txt]*:    VDJ    region length distribution plots with underlying data for the assembled BCRs for a locus.

6. *clonotype_sizes.[pdf|txt]*: Bar graph with underlying data visualizing clonotype size distribution.

7. *clonotype_network_[with|without]_identifiers.<graph_format>*: Clonotype networks with recombinant identifier strings or lines denoting the presence of recombinants.

8. *clonotype_network_[with|without]_identifiers.dot:*    Clonotype networks described in the Graphviz DOT language.

9. *lineage_trees/:* Subdirectory containing lineage trees if run with *--infer_lineage*.

10. Intermediate output files (*see* **Note 41**).

**3.4    Quality Control of Output**

A current challenge of scRNA-seq is being able to detect and filter out reads that are not in fact derived from a single cell, but rather from unintentional cell multiplets or cross-contamination due to PCR chimeras or free RNA from lysed cells [29]. The number of reconstructed chains for a locus may be used to filter out multiple captures or potential contaminations because a single B- or T-cell should not have more than two recombined antigen receptor chains for a given locus. It is important to filter out such cells from the dataset as they otherwise could hinder correct clonotype inference. Furthermore, TraCeR and BraCeR are built on the assumption that each cell contains a maximum of two reconstructed sequences for each BCR/TCR locus, and BCR/TCR reconstruction from bulk samples or unintentional cell multiplets may therefore potentially give rise to some incorrectly reconstructed sequences. Filtering of suspected cell multiplets is done automatically for BraCeR, and can be employed manually for TraCeR.

*3.4.1    Automatic Cell Multiplet Detection*

BraCeR identifies potential cell multiplets or cross-contamination if more than two recombined sequences are reconstructed for any one BCR locus in a cell. Such cells are then excluded from further analysis steps unless *summarise* was run with *--include_multiplets*.

*3.4.2    Manual Inspection of Potential Cell Multiplets*

The frequency of cell multiplets may vary from dataset to dataset, and the importance of removing potential cell multiplets versus the risk of filtering out false potential multiplets may also vary depending on the biological question and experimental setup. Filtering of potential multiplets could therefore be done with several degrees of strictness, and should be determined by the user for each individual dataset. Our general recommendations for manual inspection and removal of potential cell multiplets are (from more permissive to more restrictive filtering):

1. Run *[tb]racer summarise* with *--use_unfiltered*.

2. Create a new directory for cells to be filtered out.

3. Open the *[TB]CR_summary.txt* in the unfiltered summary folder and look at the section named "#Cells with more than two recombinants for a locus#". Take note of any cell that has more than three reconstructed sequences for any locus, and move the result folder from the assembly step for these cells to the new folder for filtered cells.

4. Open the *<cell_name>/unfiltered_[TB]CR_seqs/unfiltered_[TB]CRs.txt* file for each cell with more than two recombinants for a locus. Discard cell if all recombinants for the locus are substantially different from each other (*see* **Note 42**).

5. Look at the clonotype network using the unfiltered cells. If one cell containing multiple chains for a locus connects to two or more distinct clone groups with their own set of sequences not shared with other sub-clone-groups, the cell is likely to be a cell multiplet.

6. Depending on the desired balance of retaining potential cell multiplets versus discarding false multiplets, you could also take into account cells in which two distinct productive recombinants have been reconstructed for more than one locus in a cell, e.g., T cells with two productive TCRα in addition to two productive TCRβ or B cells with two productive IgH and also two productive Igκ and/or Igλ. Such cells have a higher probability of being cell multiplets, although discarding them may mask true biological information.

7. Cells with two productive recombinants for a locus are expected at varying frequencies (e.g., TCRα: 30%, TCRβ: 2–10%, IgH: 2–5%, Igκ: 11% in mice) [30]. If you observe significantly higher proportions in your data, you may wish to consider the likelihood that some of these represent doublets.

*3.5 Interpreting TraCeR Clonotype Output*

The clonal inference based on reconstructed TCRs is represented as graphical output with either horizontal lines indicating whether a recombinant for each locus is present (Fig. 4a) or full recombinant identifiers (Fig. 4b). The network without identifiers gives a good overview of the overall clonality in the cell population, whereas the network with identifiers only shows clonally expanded cells and details the identity of the shared TCR sequences.

The networks shown in Fig. 4 show edges between nodes representing cells that share one or more reconstructed sequences. Whether sharing of TCR sequences can be seen as evidence of clonality depends on how strictly you wish to define clonality. Given that detection sensitivity is not 100% (and depends on various experimental and biological parameters) all the TCR sequences present in a cell may not always be reconstructed. Clone groups

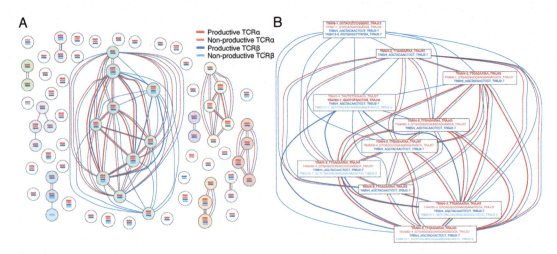

**Fig. 4** Example of a clonotype network from a mouse 14 days after infection with *Salmonella*. Each node represents a T cell (**a**). Horizontal bars represent reconstructed TCR sequences, with dark colors being productive and light colors nonproductive. Edges between nodes indicate sharing of one or more TCRs for each locus with edge thickness being proportional to the number of sequences shared between two nodes. The clone groups have different node background colors for visualization purposes. An example of a network with identifiers showing only one of the clone groups from the same mouse is shown in (**b**). Figures are adapted from Stubbington [9]

must therefore be inferred based on various levels of evidence and interpreted by the user depending on the biological questions to be answered. Here we discuss a few patterns that could be observed in the clonotypes networks, exemplified by clone groups in Fig. 4a.

1. Many small clone groups consist of cells all sharing the same productive TCRα and productive TCRβ (e.g., the green, pink, purple yellow, orange, gray and red clones).

2. Many of the clone groups also exhibit sharing of additionally reconstructed chains when these were detected (pink, turquoise, yellow, and orange clones). Sharing of such additional TCR sequences within a clone group strengthens the evidence of correct clonal assignments due to the extremely small likelihood that two independent cells would undergo the same complete set of recombination events during development in the thymus.

3. It is possible for a single TCRβ to be found in combination with multiple TCRα because developing T cells first recombine the TCRβ-locus and proliferate before recombining their TCR-α-loci. Examples of this can be seen as subclones sharing both a TCRα and TCRβ within a larger clone group of cells only sharing a TCRβ (turquoise clone). Such groups may indicate that, in these cases, the TCRβ is important in conferring antigen specificity.

4. In some cases, the only shared TCR sequence may be nonproductive or a TCRα due to failed reconstruction of other chains (e.g., blue clone). We cannot be certain that the cell belongs to the clone group, but we also have no evidence to the contrary.

5. Some strange clone groups may appear as cells all sharing different single TCR chains (e.g., blue-gray clone). This is likely not a real clone group, as all of the cells have a TCRβ and TCRα reconstructed, but some of them share only a TCRα while others share only a TCRβ, but not the same TCRβ for all the cells.

6. The presence of nodes connecting two or more smaller clone groups could be an indication of unsuccessfully removed cell multiplets (not seen in Fig. 4a).

## 3.6 Interpreting BraCeR Clonotype Output

### 3.6.1 Clonotype Networks

As for TraCeR, the clonal inference based on reconstructed BCRs is represented as graphical output with horizontal lines indicating whether a clonally related recombinant for each locus is present (Fig. 5a) or full recombinant identifiers (Fig. 5b). Here, we discuss a few patterns that could be observed in the BraCeR clonotype networks, exemplified by clone groups in Fig. 5a.

1. Unless BraCeR is run with *--IGH_networks*, all the cells in each clone group are required to share at least one productive IgH and one productive Igκ or Igλ. This requirement makes the clonal assignments fairly certain.

2. If BraCeR is run with *--IGH_networks*, larger clone groups with cells sharing a clonally related IgH may consist of subclones sharing a specific light chain. This cannot be exemplified by Fig. 5a as, in this instance, BraCeR was not run with *--IGH_networks*.

3. Sharing of additional reconstructed BCR sequences, either productive or nonproductive, within a clone group strengthens the evidence of correct clonal assignments due to the extremely small likelihood that two independent cells would undergo the same complete set of recombination events during development in the bone marrow.

4. Some cells have additionally reconstructed chains that are not shared within the clone group (e.g., one cell in the largest clone group having two productive Igλ). This could be due to varying expression levels and hence differences in reconstruction sensitivity, technical issues such as contamination or misassemblies, or display true biological variability (*see* **Note 43**).

5. Clone groups spanning different isotypes and subtypes of main isotypes may be observed (e.g., two of the clone groups spanning IgA1 and IgG1), indicating that members of the clone have undergone class-switching.

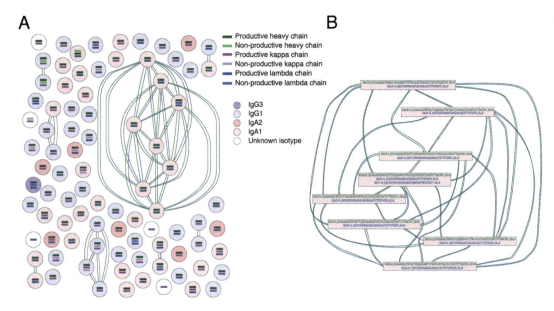

**Fig. 5** Example of a clonotype network from a human donor (**a**). Each node represents a plasmablast. Horizontal bars indicate reconstructed BCR sequences in each cell, with dark colors denoting productive and light colors nonproductive chains. Sharing of one or more clonally related BCRs for each locus is visualized as edges between the nodes, with edge thickness being proportional to the number of shared sequences. The background color of each node indicates the isotype of the cell. An example of a network with identifiers for one of the clone groups is shown in (**b**). Figure a is reproduced from Lindeman [10], and the figures are based on raw scRNA-seq data published in [14]

*3.6.2 Lineage Trees*

The lineage trees resulting from running BraCeR with *--infer_lineage* are useful to acquire more information about the similarity of the clonally related sequences within a clone group and how they may have evolved through affinity maturation. These lineage trees are built using maximum parsimony (*see* **Note 44**) with the inferred combined heavy and light chain germline sequence as outgroup (black node) and inferred intermediate sequences not observed in the sample as white nodes (Fig. 6). The larger nodes in each lineage tree are labeled with the cell name(s) containing the sequence representing each node, and the background color of each node corresponds to the isotype(s) of the IgH. The size of each node is proportional to the number of cells in which the sequence was reconstructed.

*3.6.3 Further Repertoire Analysis Using External Tools*

The output of BraCeR may be further analyzed with other available tools for BCR repertoire analysis such as the Change-O suit for analysis of SHM, lineage reconstruction, and repertoire diversity. BraCeR aims to follow common data standards as they are being outlined by the Adaptive Immune Receptor Repertoire (AIRR) community [31] in order to facilitate use of external tools (*see* **Note 45**). Most current tools for BCR repertoire analysis are designed for high-throughput repertoire sequencing data of bulk

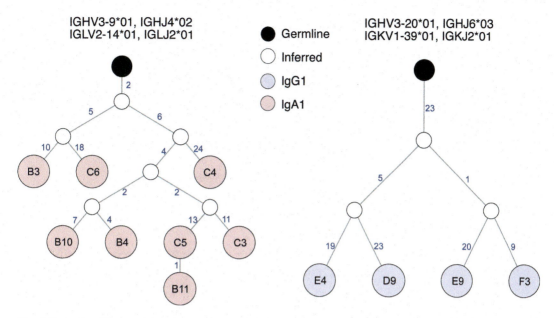

**Fig. 6** Example of lineage trees constructed for the two largest clone groups in Fig. 5. Each node represents the combined productive IgH and light chain sequence of a plasmablast. Edges between nodes indicate the edit distance between the sequences. The background color of each node indicates the isotype of the IgH. The figure is adapted from Lindeman [10]

samples, and do not automatically deal with paired heavy and light chain sequences. Some practical guidelines for BCR sequencing repertoire analysis have been reviewed in [32].

**3.7  Building Resources with Build**

*3.7.1  Introduction*

The *Build* mode of [TB]raCeR creates the required resources from user-specified reference sequences. It can be used to run [TB] raCeR for species other than human or mouse, or to use a particular set of reference sequences. The *Build* mode creates synthetic reference sequences called combinatorial recombinomes, consisting of every combination of V-alleles and J-alleles, with a masked junctional region and leader sequence to allow mapping of TCR- or BCR-derived reads to a reference. For TCRs, the first ~260 nucleotides of the constant region gene are then appended to each synthetic sequence for the locus to allow mapping of reads running into the C region (*see* **Note 46** for BCRs). *Build* also creates databases compatible with IgBLAST and BLAST.

*3.7.2  Running Build*

1. Download ungapped V, D and J reference sequences from IMGT or another repository (*see* **Note 47**).

2. If you are building resources for BraCeR, also download IMGT-gapped V reference sequences from IMGT.

3. Download constant region (C) sequences from IMGT or Ensembl (*see* **Note 48**).

4. For each locus, run one of the following commands, inserting your choices and optional arguments (*see* **Note 49**):

*tracerbuild<species><receptor_name><locus_name><N_padding> <color> <V_seqs> <J_seqs> <C_seqs> [<D_seqs>] [options] bracer build <species> <locus_name> <N_padding> <color> <V_seqs> <J_seqs> <C_seqs> [<D_seqs>] [options]*

TraCeR and BraCer both expect the following main arguments:

1. *<species>*: Species (e.g., Hsap).

2. *<locus_name>*: Name of locus (e.g., H).

3. *<N_padding>*: Number of ambiguous N nucleotides between V and J (*see* **Note 50**).

4. *<color>*: HTML color for productive recombinants (e.g., E41A1C) or *random*.

5. *<V_seqs>*: FASTA file containing V-gene sequences.

6. *<J_seqs>*: FASTA file containing J-gene sequences.

7. *<C_seqs>*: FASTA file containing a single (TraCeR) or several (BraCeR) constant region sequence(s).

8. *<D_seqs>*: FASTA file containing D-gene sequences (optional).

TraCeR and BraCeR accept the following optional arguments:

1. *-f/--force_overwrite*: Forces the program to overwrite existing resources for the species.

2. *-c/--config_file:* Path to the configuration file (*see* **Note 13**). Default = ~/.[tb]racerrc.

3. *--resource_dir:* Path to directory containing resources required for alignment. Use if you wish to use other resources contained somewhere other than the default resources directory.

4. *-o/--output_dir:* Path to write new resource files. New resources will be written to the default resource directory if not specified.

The following optional arguments are specific for BraCeR:

1. *--C_db*: Takes FASTA file containing all C-gene sequences as argument if not all C-gene sequences were used to make the combinatorial recombinomes.

2. *--V_gapped*: Takes FASTA file with IMGT-gapped V reference sequences as argument (highly recommended, *see* **Note 51**).

3. *--igblast_aux*: Takes an IgBLAST auxiliary file for the species as argument, aiding correct CDR3 assignments.

## 4    Notes

1. We use the Smart-seq2 protocol [33], with SmartScribe (Clontech) instead of SuperScript II (Invitrogen) for reverse transcription. A comparison of several available scRNA-seq protocols is presented by Svensson et al. [34]. Data generated from droplet-based sequencing such as 10× Genomics Chromium 3' Single-Cell RNA-sequencing are not suitable for TCR/BCR reconstruction by [TB]raCeR because they do not sequence full-length mRNA. However, BCR sequences obtained by the 10× Genomics Chromium Single Cell V(D)J kits may be used with the BraCeR pipeline for clonality analysis.

2. Read lengths of 50 bases or more are recommended for optimal reconstruction sensitivity and accuracy [9, 35]. If you wish to use read lengths shorter than 50 bases, you will be restricted to using the Inchworm component of Trinity (*see* **Note 13**).

3. We would generally recommend to sequence T- or B-cells with a read depth of 0.25–0.50 million PE reads per cell (in total 0.5–1.0 million reads per cell). While TCR- and BCR-sequences may be successfully reconstructed with lower sequencing depths [9, 35], the optimal depth depends on the cell type, activation state and on how much information about the rest of the transcriptomic profile for each cell is needed for downstream analyses.

4. Sequencing facilities often run quality control as part of their pipelines and let you know if there were any technical challenges during the run. Alternatively, you can run FastQC (http://www.bioinformatics.babraham.ac.uk/projects/fastqc/) on the raw reads.

5. Python (≥3.4.0) is required for BraCeR. TraCeR works with both Python 2 and Python 3.

6. BraCeR requires Trinity version 2.4.0 or higher. This is also the recommended version for TraCeR, although earlier versions may also be used.

7. Please note that in addition to downloading the executable files from ftp://ftp.ncbi.nih.gov/blast/executables/igblast/release/<version_number> you must download the *internal_data* directory and put it into the same directory as the igblast executable. Run *export IGDATA=/<path_to_igblast>/igblast/1.4.0/bin* to set the *$IGDATA* environment variable to point to the location of the igblast executable.

8. BraCeR requires Kallisto, while TraCeR may be run with either Kallisto or Salmon for quantification of the reconstructed sequences.

9. In order to visualize clonotype networks, the drawing programs Dot and Neato need to be installed.

10. PHYLIP dnapars, R and the R packages are optional, but required in order to run BraCeR with immunoglobulin lineage reconstruction. Trim Galore! is optional, but recommended to trim adapter sequences and low-quality reads before BCR reconstruction.

11. We recommend first installing numpy and biopython through your package manager or conda before setting up Python dependencies with *pip install -r requirements.txt.*

12. The resulting binaries *tracer* or *bracer* may now be run from anywhere. However, the absolute path to where [TB]raCeR was downloaded needs to be specified in the configuration file if the binary is run from outside the main [TB]raCeR directory.

13. It is important to edit the configuration file before running [TB]raCeR. An example configuration file is included in the repository (*[tb]racer.conf*). By default, this is ~/.[tb]racerrc. If [TB]raCeR fails to find this file, it will use the *[tb]racer.conf* in the repository. [TB]raCeR automatically looks in your system's PATH for the required tools. Alternatively, you can edit the *[tool_locations]* section of the configuration file to specify the path to the executables. Please make sure to also edit the [*base_transcriptomes*] (TraCeR) or [*kallisto_transcriptomes*] (BraCeR) section to include paths to location of the transcriptome FASTA file. This must be a plain-text file, please decompress it if necessary. To activate the short read mode in TraCeR (for reads shorter than 50 bases), uncomment *inchworm_only=True* and uncomment *trinity_kmer_length*. The *trinity_kmer_length* may be adjusted (17 is the minimum value, and 25 is default in Trinity). Trinity version 2 is required to run the short read mode.

14. You can pass all the usual arguments (except *--small_index*) to the Docker command, but without having to edit the configuration file.

15. [TB]raCeR may run out of memory during the assembly step if the memory limit for Docker is not increased to 6–8 GB. Instructions on how to do this can be found at https://docs.docker.com/docker-for-windows/#advanced for Windows and https://docs.docker.com/docker-for-mac/#advanced for Mac.

16. The TraCeR test data are derived from mouse T cells, while the data used in the BraCeR test are from human B cells. Make sure that the configuration file contains the correct resource files (mouse for TraCeR — human for BraCeR). Run *bracer test* with *--infer_lineage* to test your installation of the additionally required tools for lineage reconstruction. If you are running

the test on Docker, you need to clone the GitHub repository, enter its main directory, and call *docker run -it --rm -v $PWD:/ scratch -w /scratch teichlab/[tb]racer test -o test_data*.

17. Running *tracer test* should result in three cells, of which Cell 1 and Cell 2 are in a clonotype. Each cell should have a productive TCRα, a nonproductive TCRα, a productive TCRβ, and a nonproductive TCRβ. The output of *bracer test* should also be three cells, with Cell 2 and Cell 3 belonging to a clonotype. Two of the cells should have a productive IgH and a productive Igλ, and one cell should have a productive IgH, productive Igλ, and nonproductive Igκ.

18. Adapter sequences and low-quality sequences are by default trimmed from the raw reads using Trim Galore! (http://www. bioinformatics.babraham.ac.uk/projects/trim_galore/) and Cutadapt [36].

19. Bowtie 2 is run with low penalties for insertion of gaps into the read or reference sequence and mapping to ambiguous N nucleotides in the reference sequence.

20. IgH CDR3 regions are variable in length and may be relatively long. BraCeR therefore runs a second round of alignment to extract CDR3-derived reads. In this step, all reads are mapped locally to the IgH reads extracted in the first alignment step in order to identify reads that partially or fully overlap the already extracted reads, thus retrieving all the reads necessary to reconstruct the full recombinant sequence. This alignment is run with high penalties for mismatches and introduction of gaps.

21. Standalone BLAST is used to determine the C-region gene for each BCR sequence.

22. This parameter is used to filter out artifacts. The default --*max_junc_len* is 50 for TraCeR and 100 for BraCeR, but may need to be set higher for γδ T cells as TCRδ chains are known to have highly variable CDR3 lengths [37].

23. The default mode of TraCeR (*-m imgt*) is to replace all but the junctional region of the reconstructed TCRs with the most likely IMGT reference sequence before assessing whether the recombinant is productively rearranged. This method may not be suitable for humans and other outbred populations with high uncharacterized genetic variability, as it ignores non-junctional sequence changes compared with the IMGT references. In these cases it may be better to run *tracer assemble* with *-m assembly* in order to base all analyses on the reconstructed sequences for humans.

24. To use the *--small-index* option, specify the location of an index built from the corresponding base_transcriptome in the configuration file under [[*salmon|kallisto]_base_indices*]. Reads

will then first be quantified with the *base_index* before a small index is built from the expressed transcripts and reconstructed TCRs, and the reads are quantified with this small index.

25. If FASTQ file(s) are provided, BraCeR also quantifies the BCR sequences.

26. TraCeR always uses the junctional nucleotide sequence reported by IgBLAST, while BraCeR replaces this sequence with the identified CDR3 nucleotide sequence if detected.

27. The *unfiltered_[TB]CR_seqs* subdirectory contains several files. *unfiltered_[TB]CRs.txt* is a text file detailing the reconstructed TCR/BCR recombinants. *<cell_name>_[TB]CRseqs.fa* is a FASTA file listing the reconstructed TCR/BCR sequences. *<cell_name>.pkl* is a Python pickle file which is used in the *summarise* step.

28. If *assemble* is run with *--small_index*, only the quantification output using the small index will be found here.

29. The CDR3 nucleotide distance calculation is based on a human 5-mer targeting model [38], mouse 5-mer targeting model [39] or nucleotide Hamming distance for other species. Other distance threshold values can be specified with the *--dist* argument.

30. B-cell clonality is based on a shared potentially clonally related IgH and light chain to increase the confidence of clonal assignment because it can be hard to determine whether very similar chains are the result of SHM or rather similar recombination events occurring by chance during B-cell development. As shared light chain sequences are more likely to occur by chance as a result of similar recombination events compared to IgH, we do not show sharing of potentially clonally related light chains if the cells do not also share a clonally related IgH. If you wish to base clonality only on IgH, for example if you are studying developing B cells, *bracer summarise* may be run with *--IGH_networks*. Information about reconstructed light chains in the cells will then be included in the networks, but will not affect clonal clustering.

31. IMGT-gapped resources for mouse and human can be found within the resources directory, and may be generated through *Build* for any other species.

32. Run *summarise* with *--IGH_network* to draw lineage trees separately for each locus. This will include cells in IgH lineage trees even if they have completely different light chains.

33. In short, the *lineage.R* script loads a tab-delimited database file in the Change-O data format. Identical sequences within each clone group are then collapsed into one and annotated with the cell names, isotypes, and number of cells containing the

sequence. Maximum parsimony lineage trees are built through the *dnapars* method of PHYLIP with the Alakazam function *buildPhylipLineage* for each clone group. Lastly, we parse the output and modify the tree topology.

34. To remove certain cells from downstream analyses, you can move the individual result directories for each cell somewhere other than the directory used as input for the *summarise* step.

35. If you have defined new species using *Build*, you should specify the same name here.

36. TraCeR recognizes TCRα (A), TCRβ (B), TCRγ (G), and TCRδ (D) and BraCeR accepts IgH (H), Igκ (K), and Igλ (L) for mouse and human. Other locus names can be specified if you have created resources with the *Build* module. By default, TraCeR will attempt to summarize TCRα and TCRβ sequences, while BraCeR will attempt to summarize IgH, Igκ, and Igλ sequences. To change this, pass a space-delimited list of locus names. For example, to only look for TCRγ and TCRδ use *--loci G D*.

37. The output format needs to be one of the options detailed at http://www.graphviz.org/doc/info/output.html. In our experience the PDF format does not work when trying to summarize more than a few cells, and we recommend using the *svg* format for high-resolution figures.

38. Use this option if you do not have Graphviz installed.

39. TraCeR attempts to identify invariant natural killer T (iNKT) cells [40] and mucosa-associated invariant T (MAIT) cells [41] by their characteristic TCRα gene segments. These are removed before creation of networks.

40. This distance value may need to be increased for datasets containing sequences with many SHMs, or to be decreased for datasets consisting mainly of naïve B cells. The optimal distance threshold for a dataset can be determined with the SHazaM package [26]. See the developers' vignette (https://shazam.readthedocs.io/en/version-0.1.9%2D%2D-baseline-fixes/vignettes/DistToNearest-Vignette/) for a step-by-step explanation on how to do this.

41. Several intermediate output files may be generated when running BraCeR summarise. For a list and description of these files please see our documentation pages at https://github.com/Teichlab/bracer.

42. To check if recombinants for a locus are substantially distinct from each other we first inspect the V- and J-gene-usage of each recombinant. If two or more of the recombinants have the same V-gene- and J-gene-usage, they may represent the same rearranged TCR/BCR sequence. The recombinant with the

lowest expression value is in such cases likely to be a misassembled sequence. If you wish to base your filtering on more detailed analysis, we recommend aligning the similar reconstructed sequences for the locus using a pairwise alignment tool such as the EMBOSS Matcher Tool [42] in nucleotide mode to identify the differences between them. If the difference is a single nucleotide or an insertion/deletion, especially in a homopolymer tract, the sequence with the lowest expression value is likely a misassembly due to sequencing errors and/or PCR errors.

43. True biological variability could potentially be the result of secondary rearrangements such as receptor editing of autoreactive B cells [43–45]. For example, we often observe some cells with both a reconstructed productive Igκ and Igλ. Usually one of the chains is highly expressed while the other is very lowly expressed.

44. Maximum parsimony makes several assumptions that may not always hold true for immunoglobulin lineages. For the most accurate lineage tree reconstruction it could therefore be worth also employing other phylogenetic methods more specifically designed for immunoglobulin lineage reconstruction, such as IgPhyML [46].

45. BraCeR creates output databases following the Change-O data format (described in http://changeo.readthedocs.io/en/version-0.3.12%2D%2D-makedb-fix/standard.html). BraCeR also adds some additional columns to the database, as described in [10].

46. Due to immunoglobulin class switching, IgH chains may be expressed in combination with one of multiple constant (C) regions, depending on isotype. To allow for alignment of reads regardless of isotype, we appended one or a few representative sequences for each isotype (IgH) or one representative Igκ or Igλ C-region sequence to the 3' end of each entry in the appropriate synthetic recombinant files for each locus.

47. It is not necessary to parse the sequence annotations of the reference sequences provided to BraCeR, as this will be done automatically.

48. To increase performance, prepare an additional file containing only a few representative C-alleles for use with BraCeR, and pass this file as *C_seqs*. Provide the full C reference file with the optional argument *--C_db* for accurate isotype detection.

49. To run the Docker version of *[tb]racer build* you must specify *--resource_dir /scratch* in order to save the created resources. The newly created resource may then be used by running the Docker image from the same directory as used to build the resources, again specifying *--resource_dir /scratch*.

50. The resources distributed with [TB]raCeR were created using an N padding of 7 nucleotides for TCRβ and TCRδ, 8 nucleotides for IgH, and one nucleotide for TCRα, TCRγ, Igκ, and Igλ. In our experience, changing the N padding will not have a large effect on reconstruction sensitivity and accuracy.

51. Providing IMGT-gapped V reference sequences is required for lineage reconstruction and creation of IMGT-gapped tab-delimited databases. We highly recommend providing these sequences even if you are not running lineage reconstruction in order to accurately determine recombinant productivity and CDR3 regions.

# References

1. Bashford-Rogers RJM, Palser AL, Huntly BJ, Rance R, Vassiliou GS, Follows GA, Kellam P (2013) Network properties derived from deep sequencing of human B-cell receptor repertoires delineate B-cell populations. Genome Res 23(11):1874–1884. https://doi.org/10. 1101/gr.154815.113

2. Weinstein JA, Jiang N, White RA 3rd, Fisher DS, Quake SR (2009) High-throughput sequencing of the zebrafish antibody repertoire. Science 324(5928):807–810. https://doi.org/10.1126/science.1170020

3. Wang C, Sanders CM, Yang Q, Schroeder HW Jr, Wang E, Babrzadeh F, Gharizadeh B, Myers RM, Hudson JR Jr, Davis RW, Han J (2010) High throughput sequencing reveals a complex pattern of dynamic interrelationships among human T cell subsets. Proc Natl Acad Sci U S A 107(4):1518–1523. https://doi.org/10. 1073/pnas.0913939107

4. Rosati E, Dowds CM, Liaskou E, Henriksen EKK, Karlsen TH, Franke A (2017) Overview of methodologies for T-cell receptor repertoire analysis. BMC Biotechnol 17:61. https://doi. org/10.1186/s12896-017-0379-9

5. Bashford-Rogers RJM, Palser AL, Idris SF, Carter L, Epstein M, Callard RE, Douek DC, Vassiliou GS, Follows GA, Hubank M, Kellam P (2014) Capturing needles in haystacks: a comparison of B-cell receptor sequencing methods. BMC Immunol 15:29. https://doi. org/10.1186/s12865-014-0029-0

6. Han A, Glanville J, Hansmann L, Davis MM (2014) Linking T-cell receptor sequence to functional phenotype at the single-cell level. Nat Biotechnol 32:684. https://doi.org/10. 1038/nbt.2938

7. Kolodziejczyk AA, Lönnberg T (2017) Global and targeted approaches to single-cell transcriptome characterization. Brief Funct Genom 17:209–219. https://doi.org/10. 1093/bfgp/elx025

8. Stubbington MJT, Rozenblatt-Rosen O, Regev A, Teichmann SA (2017) Single-cell transcriptomics to explore the immune system in health and disease. Science 358 (6359):58–63. https://doi.org/10.1126/science.aan6828

9. Stubbington MJT, Lonnberg T, Proserpio V, Clare S, Speak AO, Dougan G, Teichmann SA (2016) T cell fate and clonality inference from single-cell transcriptomes. Nat Methods 13 (4):329–332. https://doi.org/10.1038/nmeth.3800

10. Lindeman I, Emerton G, Mamanova L, Snir O, Polanski K, Qiao SW et al (2018) BraCeR: B-cell-receptor reconstruction and clonality inference from single-cell RNA-seq. Nat Methods 15(8):563–565

11. Eltahla AA, Rizzetto S, Pirozyan MR, Betz-Stablein BD, Venturi V, Kedzierska K, Lloyd AR, Bull RA, Luciani F (2016) Linking the T cell receptor to the single cell transcriptome in antigen-specific human T cells. Immunol Cell Biol 94(6):604–611

12. Rizzetto S, Koppstein DNP, Samir J, Singh M, Reed JH, Cai CH et al (2018) B-cell receptor reconstruction from single-cell RNA-seq with VDJPuzzle. Bioinformatics 34 (16):2846–2847

13. Afik S, Yates KB, Bi K, Darko S, Godec J, Gerdemann U, Swadling L, Douek DC, Klenerman P, Barnes EJ, Sharpe AH, Haining WN, Yosef N (2017) Targeted reconstruction of T cell receptor sequence from single cell RNA-seq links CDR3 length to T cell differentiation state. Nucleic Acids Res 45(16):e148. https://doi.org/10.1093/nar/gkx615

14. Canzar S, Neu KE, Tang Q, Wilson PC, Khan AA (2017) BASIC: BCR assembly from single

cells. Bioinformatics 33(3):425–427. https://doi.org/10.1093/bioinformatics/btw631

15. Upadhyay AA, Kauffman RC, Wolabaugh AN, Cho A, Patel NB, Reiss SM, Havenar-Daughton C, Dawoud RA, Tharp GK, Sanz I, Pulendran B, Crotty S, Lee FE-H, Wrammert J, Bosinger SE (2018) BALDR: a computational pipeline for paired heavy and light chain immunoglobulin reconstruction in single-cell RNA-seq data. Genome Med 10 (1):20. https://doi.org/10.1186/s13073-018-0528-3

16. Lönnberg T, Svensson V, James KR, Fernandez-Ruiz D, Sebina I, Montandon R, Soon MSF, Fogg LG, Nair AS, Liligeto UN, Stubbington MJT, Ly L-H, Bagger FO, Zwiessele M, Lawrence ND, Souza-Fonseca-Guimaraes F, Bunn PT, Engwerda CR, Heath WR, Billker O, Stegle O, Haque A, Teichmann SA (2017) Single-cell RNA-seq and computational analysis using temporal mixture modeling resolves TH1/TFH fate bifurcation in malaria. Sci Immunol 2(9). https://doi.org/10.1126/sciimmunol.aal2192

17. Patil VS, Madrigal A, Schmiedel BJ, Clarke J, O'Rourke P, de Silva AD, Harris E, Peters B, Seumois G, Weiskopf D, Sette A, Vijayanand P (2018) Precursors of human CD4(+) cytotoxic T lymphocytes identified by single-cell transcriptome analysis. Sci Immunol 3(19). https://doi.org/10.1126/sciimmunol.aan8664

18. Miragaia RJ, Gomes T, Chomka A, Jardine L, Riedel A, Hegazy AN, Lindeman I, Emerton G, Krausgruber T, Shields J, Haniffa M, Powrie F, Teichmann SA (2017) Single cell transcriptomics of regulatory T cells reveals trajectories of tissue adaptation. bioRxiv. https://doi.org/10.1101/217489

19. Langmead B, Salzberg SL (2012) Fast gapped-read alignment with bowtie 2. Nat Methods 9 (4):357–359. https://doi.org/10.1038/nmeth.1923

20. Grabherr MG, Haas BJ, Yassour M, Levin JZ, Thompson DA, Amit I, Adiconis X, Fan L, Raychowdhury R, Zeng Q, Chen Z, Mauceli E, Hacohen N, Gnirke A, Rhind N, di Palma F, Birren BW, Nusbaum C, Lindblad-Toh K, Friedman N, Regev A (2011) Full-length transcriptome assembly from RNA-Seq data without a reference genome. Nat Biotechnol 29(7):644–652. https://doi.org/10.1038/nbt.1883

21. Ye J, Ma N, Madden TL, Ostell JM (2013) IgBLAST: an immunoglobulin variable domain sequence analysis tool. Nucleic Acids Res 41(Web Server issue):W34–W40. https://doi.org/10.1093/nar/gkt382

22. Bray NL, Pimentel H, Melsted P, Pachter L (2016) Near-optimal probabilistic RNA-seq quantification. Nat Biotechnol 34 (5):525–527. https://doi.org/10.1038/nbt.3519

23. Patro R, Duggal G, Love MI, Irizarry RA, Kingsford C (2017) Salmon provides fast and bias-aware quantification of transcript expression. Nat Methods 14:417. https://doi.org/10.1038/nmeth.4197

24. Camacho C, Coulouris G, Avagyan V, Ma N, Papadopoulos J, Bealer K, Madden TL (2009) BLAST+: architecture and applications. BMC bioinformatics 10:421. https://doi.org/10.1186/1471-2105-10-421

25. Ilicic T, Kim JK, Kolodziejczyk AA, Bagger FO, McCarthy DJ, Marioni JC, Teichmann SA (2016) Classification of low quality cells from single-cell RNA-seq data. Genome Biol 17(1):29. https://doi.org/10.1186/s13059-016-0888-1

26. Gupta NT, Vander Heiden JA, Uduman M, Gadala-Maria D, Yaari G, Kleinstein SH (2015) Change-O: a toolkit for analyzing large-scale B cell immunoglobulin repertoire sequencing data. Bioinformatics 31 (20):3356–3358. https://doi.org/10.1093/bioinformatics/btv359

27. Stern JN, Yaari G, Vander Heiden JA, Church G, Donahue WF, Hintzen RQ, Huttner AJ, Laman JD, Nagra RM, Nylander A, Pitt D, Ramanan S, Siddiqui BA, Vigneault F, Kleinstein SH, Hafler DA, O'Connor KC (2014) B cells populating the multiple sclerosis brain mature in the draining cervical lymph nodes. Sci Transl Med 6(248):248ra107. https://doi.org/10.1126/scitranslmed.3008879

28. Felsenstein J (1989) PHYLIP - phylogeny inference package (version 3.2). Cladistics 5:164–166 doi:citeulike-article-id:2344765

29. Goldstein LD, Chen Y-JJ, Dunne J, Mir A, Hubschle H, Guillory J, Yuan W, Zhang J, Stinson J, Jaiswal B, Pahuja KB, Mann I, Schaal T, Chan L, Anandakrishnan S, Lin C-w, Espinoza P, Husain S, Shapiro H, Swaminathan K, Wei S, Srinivasan M, Seshagiri S, Modrusan Z (2017) Massively parallel nanowell-based single-cell gene expression profiling. BMC Genomics 18(1):519. https://doi.org/10.1186/s12864-017-3893-1

30. Brady BL, Steinel NC, Bassing CH (2010) Antigen receptor allelic exclusion: an update and reappraisal. J Immunol 185 (7):3801–3808. https://doi.org/10.4049/jimmunol.1001158

31. Breden F, Luning Prak ET, Peters B, Rubelt F, Schramm CA, Busse CE, Vander Heiden JA,

Christley S, Bukhari SAC, Thorogood A, Matsen Iv FA, Wine Y, Laserson U, Klatzmann D, Douek DC, Lefranc MP, Collins AM, Bubela T, Kleinstein SH, Watson CT, Cowell LG, Scott JK, Kepler TB (2017) Reproducibility and reuse of adaptive immune receptor repertoire data. Front Immunol 8:1418. https://doi.org/10.3389/fimmu.2017.01418

32. Yaari G, Kleinstein SH (2015) Practical guidelines for B-cell receptor repertoire sequencing analysis. Genome Med 7(1):121. https://doi.org/10.1186/s13073-015-0243-2

33. Picelli S, Faridani OR, Bjorklund AK, Winberg G, Sagasser S, Sandberg R (2014) Full-length RNA-seq from single cells using smart-seq2. Nat Protoc 9(1):171–181. https://doi.org/10.1038/nprot.2014.006

34. Svensson V, Natarajan KN, Ly L-H, Miragaia RJ, Labalette C, Macaulay IC, Cvejic A, Teichmann SA (2017) Power analysis of single-cell RNA-sequencing experiments. Nat Methods 14:381. https://doi.org/10.1038/nmeth.4220

35. Rizzetto S, Eltahla AA, Lin P, Bull R, Lloyd AR, Ho JWK, Venturi V, Luciani F (2017) Impact of sequencing depth and read length on single cell RNA sequencing data of T cells. Sci Rep 7(1):12781. https://doi.org/10.1038/s41598-017-12989-x

36. Martin M (2011) Cutadapt removes adapter sequences from high-throughput sequencing reads. EMBnetjournal 17(1):10–12. https://doi.org/10.14806/ej.17.1.200

37. Rock EP, Sibbald PR, Davis MM, Chien YH (1994) CDR3 length in antigen-specific immune receptors. J Exp Med 179(1):323–328. https://doi.org/10.1084/jem.179.1.323

38. Yaari G, Vander Heiden JA, Uduman M, Gadala-Maria D, Gupta N, Stern JNH, O'Connor KC, Hafler DA, Laserson U, Vigneault F, Kleinstein SH (2013) Models of somatic hypermutation targeting and substitution based on synonymous mutations from high-throughput immunoglobulin sequencing data. Front Immunol 4:358. https://doi.org/10.3389/fimmu.2013.00358

39. Cui A, Di Niro R, Vander Heiden JA, Briggs AW, Adams K, Gilbert T, O'Connor KC, Vigneault F, Shlomchik MJ, Kleinstein SH (2016) A model of somatic hypermutation targeting in mice based on high-throughput ig sequencing data. J Immunol 197(9):3566–3574. https://doi.org/10.4049/jimmunol.1502263

40. Brennan PJ, Brigl M, Brenner MB (2013) Invariant natural killer T cells: an innate activation scheme linked to diverse effector functions. Nat Rev Immunol 13(2):101–117. https://doi.org/10.1038/nri3369

41. Dias J, Leeansyah E, Sandberg JK (2017) Multiple layers of heterogeneity and subset diversity in human MAIT cell responses to distinct microorganisms and to innate cytokines. Proc Natl Acad Sci 114(27):E5434–E5443. https://doi.org/10.1073/pnas.1705759114

42. Li W, Cowley A, Uludag M, Gur T, McWilliam H, Squizzato S, Park YM, Buso N, Lopez R (2015) The EMBL-EBI bioinformatics web and programmatic tools framework. Nucleic Acids Res 43(W1):W580–W584. https://doi.org/10.1093/nar/gkv279

43. Liu S, Velez M-G, Humann J, Rowland S, Conrad FJ, Halverson R, Torres RM, Pelanda R (2005) Receptor editing can lead to allelic inclusion and development of B cells that retain antibodies reacting with high avidity autoantigens. J Immunol 175(8):5067–5076. https://doi.org/10.4049/jimmunol.175.8.5067

44. Lang J, Ota T, Kelly M, Strauch P, Freed BM, Torres RM, Nemazee D, Pelanda R (2016) Receptor editing and genetic variability in human autoreactive B cells. J Exp Med 213(1):93–108. https://doi.org/10.1084/jem.20151039

45. Pelanda R (2014) Dual immunoglobulin light chain B cells: Trojan horses of autoimmunity? Curr Opin Immunol 27:53–59. https://doi.org/10.1016/j.coi.2014.01.012

46. Hoehn KB, Lunter G, Pybus OG (2017) A phylogenetic codon substitution model for antibody lineages. Genetics 206(1):417–427. https://doi.org/10.1534/genetics.116.196303

# Chapter 16

# A Hidden Markov Random Field Model for Detecting Domain Organizations from Spatial Transcriptomic Data

## Qian Zhu

## Abstract

Cells in complex tissues are organized by distinct microenvironments and anatomical structures. This spatial environment of cells is thought to be important for division of labor and other specialized functions of tissues. Recently developed spatial transcriptomic technologies enable the quantification of expression of hundreds of genes while accounting for cells' spatial coordinates, providing an opportunity to study spatially organized structures. Here, we describe a computational pipeline for detecting the spatial organization of cells based on a hidden Markov random field model. We illustrate this pipeline with data generated from multiplexed smFISH from the adult mouse visual cortex.

**Key words** Hidden Markov random field, Spatial organization, Sequential fluorescence in situ hybridization, Multiplexed fluorescence in situ hybridization

## 1 Introduction

Determining cell types has been a crucial goal of single-cell transcriptomic profiling [1–4]. However, often within cell types, there is also cellular heterogeneity that is attributed to the cells' distinct microenvironment and spatial context, which is often not well understood. The spatial environments come in various forms and scales in mammalian tissues. For example, the mammalian liver is divided into multiple lobules along the portal field where the metabolic tasks of the tissue are spatially assigned [5]. In the brain, cells are spatially distributed across various anatomical structures, or spatial domains, each of which is uniquely associated with distinct cognitive functions and behavior [6]. Understanding the principle of spatial division of cells requires spatially profiling technologies. Recently, development of single-molecule fluorescence in situ hybridization (smFISH) [7–11] has enabled the detection of mRNA transcripts while maintaining the spatial context. In contrast to scRNAseq studies which require cells to be dissociated from their physical context, the smFISH technique faithfully captures

Guo-Cheng Yuan (ed.), *Computational Methods for Single-Cell Data Analysis*, Methods in Molecular Biology, vol. 1935, https://doi.org/10.1007/978-1-4939-9057-3_16, © Springer Science+Business Media, LLC, part of Springer Nature 2019

cells' spatial coordinates by imaging. Furthermore, when smFISH is combined with sequential rounds of imaging (seqFISH [8, 11], MERFISH [9]), it is now possible to profile hundreds of genes for each cell while recording each cell's coordinates. With spatial genomics technologies, we can now begin to examine distinct microenvironment niches and cells' interactions within these microenvironments.

One critical question that can now be answered with spatial genomics data is to determine the spatial gene expression domains [11, 12]. Conventional methods, such as K-means, cluster cells solely on expression and ignore the spatial relationships between cells. Thus simply overlaying the cluster annotations on cells' coordinates results in a noisy, inaccurate representation of spatial domain structure. Recently, we have developed an approach based on hidden Markov random field (HMRF) to dissect the spatial domain structure [12]. The approach balances intrinsic cellular expression and extrinsic neighborhood effects to probabilistically assign domain states to single cells. In this chapter, we describe a pipeline for performing a HMRF analysis using a mouse brain data set as an illustrating example.

## 2  Materials

### 2.1  Prerequisites

We require R version 3 and Python 2.7. The following Python prerequisite packages need to be installed: seaborn (0.7.0 or up), pandas, numpy, scipy, matplotlib. Use pip --user --install <package name> to install any missing packages. The following R packages are also required: lattice, misc3d, oro.nifti, pracma, Matrix, mvtnorm. We require JAVA (version 7 or 8) and GraphColoring package.

### 2.2  R smfishHmrf Package

Obtain and install the smfishHmrf R package:

```
install.packages("devtools")
library(devtools)
install_bitbucket("qzhudfci/smfishhmrf-r", ref="default")
```

### 2.3  Python smfishHmrf Package

This contains the wrapper and interface functions for interacting with the R smfishHmrf package and is required for running HMRF and downstream visualizations. Install by:

```
pip install --user smfishHmrf
```

If the smfishHmrf package has been previously installed, one can update to the latest version by:

```
pip install --user --upgrade --no-cache-dir --no-deps smfishHmrf
```

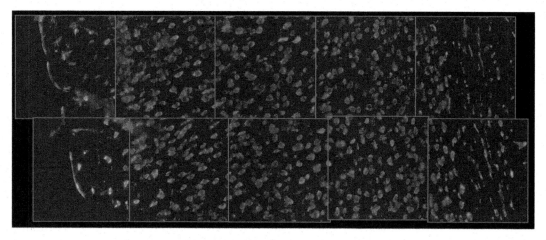

**Fig. 1** Example of imaged sections in the mouse visual cortex. Each blue box shows a section, also called a field. Sections are stitched together to form a global tissue view

### 2.4 Spatial Single-Cell Data Set

Our pipeline is general to all types of spatial transcriptomic data. For this chapter, we focus on a mouse visual cortex data set generated by the seqFISH technology. This input data set is a mouse coronal brain slice that has been imaged and is composed of various sections of the hippocampus and visual cortex tissues (Fig. 1). Each section, also called field, measures $1020 \times 1020$ units (each unit is equivalent to 220 nm). These sections are imaged and processed together with two pieces of information provided: (1) cell coordinates (two versions: relative to each field, and relative to a stitched image which stitches adjacent fields in cortex), and (2) cellular gene expression for 125 genes.

In general, we require the cell coordinate file and the gene expression file to carry out a spatial domain inference analysis. The specifications of these two files are as follows.

The cell coordinate file is made up of 4 headerless columns, separated by space: *cell_index, field_ID, x-coord, y-coord*. All fields must be numerical. Cell index $= 1 \ldots N$ where $N$ is the number of cells. Field ID specifies the field of view the cell is located in. X-coord, y-coord specify the coordinates of the cell in the respective field of view. Coordinates can be floating point decimals and can be negative. See a snippet of this file below:

```
1 0 675.080 -37.330
2 0 265.760 -231.140
3 0 753.460 -261.140
4 0 290.480 -261.520
5 0 991.430 -482.350
6 0 926.420 -675.880
7 0 414.500 -688.670
```

```
8  0  607.180  -773.680
9  0  715.720  -822.110
10 0  654.580  -896.760
11 0  472.450  -952.830
12 0  257.120  -133.350
13 0  700.010  -169.050
14 0  415.630  -252.450
```

The gene expression file is a space-separated matrix with rows being cells and the columns being genes. The order of cells (or rows) must be consistent with the order of cells in the coordinate file. Similarly, genes (or columns) are arranged in the order specified by a separate gene order file (genes.txt). First column is the row name, i.e., the cell index (equal to $1, 2, \ldots, N$). Note that there is no column header, so the first cell starts at the first line of the file. A snippet is shown below:

```
1 1.08 0.60 0.95 0.51 -1.67 0.65 1.14 0.79 0.61 0.18 0.74 0.62
0.86 0.61 0.59 0.34 -2.15 0.20 0.87 0.68 0.00 0.21 -0.00 0.49
-0.43 1.07 0.55 -1.32 0.10 1.31 1.28 0.18 0.63 0.03 0.52 0.47 0.55
0.98 0.33 0.64 -0.18 1.21 1.67 -0.37 1.04 0.11 -0.63 0.89 -0.39
-2.28 -0.02 0.66 0.92 -0.81 -0.39 -1.02 0.61 -0.05 -0.01 0.61
-0.01 -0.38 -0.05 1.52 -1.25 0.13 0.38 0.70 -0.02 0.11 -0.49 -0.35
-1.47 0.30 0.13 0.39 0.45 -2.69 0.14 0.74 0.24 -0.60 -1.08 -0.65
0.70 0.03 -1.56 -0.01 0.30 0.28 1.55 -0.25 -0.33 -0.10 -1.55 0.06
-0.77 0.41 -0.75 0.17 -1.35 -0.65 -2.06 1.42 -0.73 -2.55 -1.04
-1.35 -0.71 -0.76 -1.01 -1.69 -1.70 3.26 0.44 -1.65 -0.72 -1.40
-1.27 0.83 -0.74 -0.70 -0.05 2.65 1.76
2 1.82 1.60 2.38 -0.02 -0.26 1.94 0.08 0.07 0.91 -0.20 1.57 -0.08
0.06 1.53 0.40 0.18 0.41 1.44 0.19 0.01 0.19 -0.17 -0.26 0.06
1.18 0.02 0.01 0.06 -0.31 -0.05 -0.64 -0.01 -0.17 -0.58 1.31 0.94
0.59 1.58 -0.01 -0.10 -0.13 0.05 -0.20 0.00 -0.07 -0.07 0.19
-0.23 -0.02 -0.35 -0.68 1.27 -0.21 -0.23 -0.49 0.87 -0.33 0.41
1.25 -0.15 -0.70 -0.29 -0.47 1.98 0.01 -0.19 0.06 0.96 -0.23
-0.21 -0.19 -0.39 -0.85 -0.25 0.06 -0.40 -0.63 -0.30 -0.27 -0.29
-0.61 -0.08 -0.61 0.28 0.30 0.03 1.02 0.22 1.27 1.13 2.08 0.18
-0.22 1.31 -0.17 0.82 -0.15 1.74 -0.40 -0.40 -0.23 -2.36 -1.55
0.06 -0.97 -1.79 -1.38 -0.95 -1.69 -1.91 -2.38 -2.60 -1.49 2.32
0.16 0.05 -1.41 -2.14 -1.38 0.23 -2.20 -1.60 -1.40 2.42 0.54
```

Note that the gene expression values above have undergone (1) log-transformations, and (2) gene- and cell-wise z-scoring, from the raw mRNA counts. Thus, the values are expression z-scores and are approximately normally distributed. We recommend this normalization for our spatial transcriptomic data set, but it also works on other data sets such as MERFISH [9], tissue microarray technologies [13].

## 3    Methods

### 3.1    Background: Hidden Markov Random Field Model

We recently developed a hidden Markov random field (HMRF) method to detect major spatial patterns of expression from single cell data. HMRF is a probabilistic method for pattern recognition [14–16]. Given a user input of $K$ states, the technique classifies each cell as belonging to one of the $K$ states based on gene expression and spatially neighboring cells.

Briefly, let $S = \{s_i\}$ represent the cells in the image domain. Let $\{N_i\}$ be the local neighborhood graph which defines the nodes that are neighbors of each other. Every cell is associated with expression value $x_i : i = 1 \ldots 125$. Let $C$ be a classification function $C = \{c_i$, for each cell $s_i\}$. Overall, the posterior probability of class assignment is given by:

$$P(c_i|s_i, x_i, C_{N_i}) = 1/Z \, P(x_i|c_i, \Theta)P(c_i|s_i, C_{N_i}) \qquad (1)$$

where $P(c_i|s_i, C_{N_i})$ defines the conditional probability given the class configuration of the neighbors $C_{N_i}$. The term $P(x_i|\, c_i, \Theta)$ is the probability of observing the expression $xi$ given the class $c_i$'s Gaussian distribution function $\Theta$. Detailed mathematical explanations can be found in the original paper [12]. It suffices to say that $P(c_i|s_i, C_{N_i})$ models the extrinsic influence, or the environment made up of $s_i$'s surrounding cells, while $P(x_i|\, c_i, \Theta)$ is the intrinsic component (i.e., assigns probability based on the expression identity $x_i$). The energy function that is used to model $P(c_i|s_i, C_{N_i})$ has the Pots' model, where the energy potential V is the sum of compatible pairwise interactions in the node $s_i$'s immediate neighborhood:

$$P(c_i|s_i, C_{N_i}) = \frac{1}{Z}\exp(\beta V) = \frac{1}{Z}\exp\left(\beta\sum_{s_j \in N_i} I(c_j = c_i)\right) \qquad (2)$$

$\beta$ is a weighting constant. Parameters of the model including those in $P(c_i|s_i, C_{N_i})$ and $P(x_i|\, c_i, \Theta)$ are jointly estimated by the expectation maximization procedure.

### 3.1.1    Gene Selection

Selection of spatially coherent genes can aid the HMRF modeling. To help us find spatially coherent genes, we define a score as follows. For each gene we divide all the cells based on its bias-corrected gene expression into two classes: 1—expressed class which corresponds to 90th percentile in the gene's expression distribution, and 0—the remaining cells. We use silhouette metric to measure how spatially coherent is the 1-marked cells. Here, we used the rank-normalized, exponentially transformed distance to emphasize the local physical distance between cells. For a pair of cells, $s_i$ and $s_j$, this distance is defined as:

$$r(s_i, s_j) = 1 - p^{\mathrm{rank}_d(s_i, s_j) - 1} \qquad (3)$$

where $rank_d(s_i, s_j)$ is the mutual rank [17] of $s_i$ and $s_j$ in the vectors of Euclidean distances $\{Euc(s_i, *)\}$ and $\{Euc(s_j, *)\}$. $p$ is a rank-weighting constant set between 0.95 and 0.99. The spatial coherence of the gene is calculated as the Silhouette coefficient of the spatial distance between the two cell sets:

$$S_g = 1/|L_1|\sum\nolimits_{s_i \in L_1}(m_i - n_i)/\max(m_i, n_i) \qquad (4)$$

where for a given cell $s_i$ in $L_1, m_i$, is the average distance between $s_i$ and any cell in $L_0$, and $n_i$ is defined as the average distance between $s_i$ and any other cell in $L_1$.

*3.1.2 Neighborhood Graph Construction*

The spatial relationship between the cells is represented as an undirected graph, where each node is a cell and each edge connects a pair of neighboring cells if they are no further than a certain Euclidean distance apart. This distance threshold is a parameter of the graph construction function. We recommend setting the distance such that on average there are around 5–10 neighbors per cell.

*3.1.3 Number of Clusters*

We initialize HMRF according to the $K$-means clustering results. The value of $K$ is selected on the basis of gap-statistics.

**3.2 Application: Mouse Visual Cortex Data Set**

1. We first import the relevant packages.

```
import sys
import math
import os
import numpy as np
import scipy
import scipy.stats
import pandas as pd
from scipy.stats import zscore
from scipy.spatial.distance import euclidean,squareform,pdist
import smfishHmrf.reader as reader
from smfishHmrf.HMRFInstance import HMRFInstance
from smfishHmrf.DatasetMatrix import DatasetMatrix,
DatasetMatrixSingleField
from smfishHmrf.bias_correction import calc_bias_moving,
do_pca, plot_pca
import smfishHmrf.visualize as visualize
import smfishHmrf.spatial as spatial
```

2. Illumination bias correction is recommended for this data set. Illumination bias refers to biases arising from the optical instrument which may produce images with skewed intensity in certain regions of the imaged field. We systematically detect and remove such imaging bias using a multiple field,

smoothing average approach. The idea starts by realizing that the tissue cannot be imaged all in one time, but rather it is imaged in small subsections (fields). By comparing the intensity of the same spot across multiple fields, we can deduce an average bias for that spot for correction. This approach computes an average field bias vector by first overlaying all field images and apportioning cells into bins on a regular grid. The bias for a bin $b$ and a gene $g$ is the total expression of cells in $b$ and the four neighboring bins of $b$ across all images, divided by the number of cells in those bins, producing a smoothed estimate.

```
expr_bias = calc_bias_moving(expr=expr, position=Xcen,
field=field, interest_field=FDs, num_bin=50)
```

We next model the bias pattern of all genes using PCA (do_pca function). The contributions of the top principal components are subtracted from the expression matrix (expr). The result is contained in corrected_expr.

```
corrected_expr = do_pca(expr=expr, expr_bias=expr_bias,
centroid=Xcen, field=field, interest_field=FDs, num_bin=50,
top_comp_remove=5)
```

We plot the top PCs representing the orthogonal bias patterns (Fig. 2).

```
plot_pca(expr_bias=expr_bias, num_bin=50, top_pc=5,
out_file="bias_component.png")
```

Note that if the expression matrix is already corrected for this bias, skip this step and go to **step 3**.

3. Load the bias corrected gene expression matrix, and the relevant coordinate file. To save time, we provide the bias-corrected expression and coordinate files in a URL for download    https://bitbucket.org/qzhudfci/smfishhmrf-py/src/

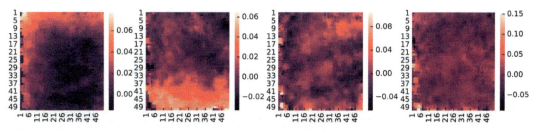

**Fig. 2** Top four principal components associated with the illumination bias. Each panel is a 50 bins by 50 bins heatmap showing the bias level associated with each region of the microscope field of view. Left to right: PC1, 2, 3, 4

default/data/. Download the content in a directory. Change the variable directory to point to the directory containing these files.

```
directory = "workdir"
genes = reader.read_genes("%s/genes" % directory)
Xcen, field =
reader.read_coordinates("%s/fcortex.coordinates.txt" % directory)
expr =
reader.read_expression_matrix("%s/fcortex.expression.txt" %
directory)
```

The expression matrix (expr, a 2D numpy array) consists of cells from the visual cortex region. Xcen and field are encoded as numpy 2D array and numpy 1D array respectively.

4. Construct a DatasetMatrixSingleField instance. DatasetMatrix-SingleField is a class that encapsulates all information about a spatial transcriptomic data set. We initiate an instance with expression, gene names, cell annotations (set to None) and coordinates (Xcen).

```
this_dset = DatasetMatrixSingleField(expr, genes, None, Xcen)
```

5. Construct a local neighborhood graph. We compute the cell pairwise Euclidean distance matrix using the cell coordinates in Xcen. Then we determine a cutoff on Euclidean distance such that a pair of cells separated by less than the cut-off distance is assigned an edge. This cutoff can be expressed as top X-percentile of Euclidean distances. For example, 0.30 means a cutoff that is equal to 0.30% of all Euclidean distance values. In this example, we settle on 0.30% as this cutoff produces on average 5 neighbors per cell (*see* output in **Note 1**). Use test_adjacency_list to test a number of cut-off values (*see* output in **Note 1**).

```
this_dset.test_adjacency_list([0.3, 0.5, 1], metric="euclidean")
this_dset.calc_neighbor_graph(0.3, metric="euclidean")
```

6. We compute the independent regions of the graph (*see* **Note 2**)—this step is required for the node update during the iterative HMRF parameter estimation.

```
this_dset.calc_independent_region()
```

7. Spatial gene selection. As a sanity check, we are interested in knowing which genes might have a spatial pattern. A spatial coherence score has been defined in Eq. 4. We first calculate a dissimilarity matrix based on the rank, exponential transformed Euclidean distance matrix between all cells' coordinates.

This gives more emphasis in the distances between cells that are close to each other (Eq. 4).

```
euc = squareform(pdist(Xcen, metric="euclidean"))
dissim = spatial.rank_transform_matrix(euc, reverse=False,
rbp_p=0.95)
```

Then we perform the silhouette calculation. Essentially, for each gene, expressions are divided into 1's (expressed) or 0's (not expressed). Silhouette metric is calculated to assess the spatial distribution for the expressed cell group (*see* **Note 3** for advanced usage of spatial.calc_silhouette_per_gene). A statistical *P*-value is reported per gene.

```
res = spatial.calc_silhouette_per_gene(genes=genes, expr=expr,
dissim=dissim, examine_top=0.1, permutation_test=True,
permutations=100)
print "gene", "sil.score", "p-value"
for i,j,k in res:
    print i,j,k
```

See the silhouette score output in **Note 4**. We select a *P* value cutoff in choosing spatial genes, to give about 90 genes. Though, users can impose additional gene restriction criterion, which we did in this example (*see* next step for additional restriction).

```
res_df = pd.DataFrame(res, columns=["gene", "sil.score", "pval"])
res_df = res_df.set_index("gene")
new_genes = res_df[res_df.pval<=0.05].index.get_values().tolist()
print new_genes
```

Some examples of spatial genes are below and we can visualize them using the following code (*see* output in Fig. 3).

```
for g in ["calb1", "acta2", "tbr1"]:
 visualize.gene_expression(this_dset, goi=g, vmax=2.0, vmin=0,
\
 title=True, colormap="Reds", size_factor=5, dot_size=20, \
 outfile="%s.png" % g)
```

8. Optionally, users may further remove cell-type-specific genes from the spatial gene list. This is particularly helpful if certain cell types are not known to form any spatial patterns, such as astrocytes and microglia, or if users wish to remove cell-type variations so as to focus solely on spatial variation in the data. In our case, we have determined a list of cell-type-specific genes strongly associated with the 8 major cell types in the cortex (based on the Tasic et al. scRNAseq data [18]) (*see* **Note 5**),

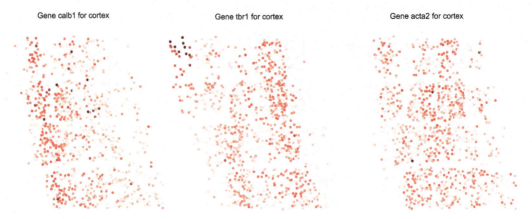

**Fig. 3** Spatial gene expression of a few spatially selected genes

and removed these genes from the spatial gene list from the previous step. This results in a 69-gene list that is used for HMRF (see file HMRF.genes).

Optionally, users may also select genes based on which genes are correlated to top principal components from PCA analysis. See the spatial.pc_genes() function, which computes the significant genes of each component with the jacksaw algorithm [19]. Users can combine this criterion and the spatial criterion in defining suitable genes for HMRF.

9. Once we decide on the genes to use for HMRF, we load this list (HMRF.genes file in the source data package), and we create a data subset using these genes with the subset_genes function.

```
new_genes = reader.read_genes("%s/HMRF.genes" % directory)
new_dset = this_dset.subset_genes(new_genes)
```

10. Initiate HMRF instance. We create an output directory. Then we construct an instance of the HMRFInstance class, which is a special class encapsulating all information about a HMRF analysis. Constructor of HMRFInstance requires *run_name output_directory DatasetMatrix_instance K (initial_beta, beta_increment, number_of_betas)*. This construction is designed to iterate HMRF over many betas.

```
print "Running HMRF..."
outdir = "spatial.jul20"
if not os.path.isdir(outdir):
 os.mkdir(outdir)
this_hmrf = HMRFInstance("cortex", outdir, new_dset, 9, (0, 0.5,
30), tolerance=1e-20)
```

As the above shows, HMRF is set to run for 30 times, starting at beta = 0, and at 0.5 increment (this covers beta = 0 up to and excluding beta = 15.0). Repeatedly running HMRF helps users see the changes in the spatial domain structure as the smoothing parameter beta is increased. Because at this stage it is uncertain what is the best beta, we need to assess all betas. Then, selecting the best beta depends on the actual data set and how the spatial pattern looks, and we discuss a beta selection guideline in **step 14**.

11. Run HMRF.

```
this_hmrf.init(nstart=1000, seed=-1)
this_hmrf.run()
```

Init() and run() are wrapper functions for R scripts where the core of HMRF is implemented. Init() determines initial conditions by running K-means to determine cluster centroids and covariance matrices, and initializing HMRF to these settings. Within init(), nstart is the number of random starts (a parameter of kmeans in R); seed allows the initial configuration to be fixed (default is −1 or unset). Run() performs the HMRF modeling, including an expectation-maximization procedure to iteratively estimate the parameters of the HMRF model, until convergence criteria is met. Run() will iterate over all K's and all beta's. At the end, files will be automatically generated in the output directory (*see* **Note 6**), with a copy of the results loaded in the class instance.

12. Next, we visualize the spatial clusters in 2D. Figure 4 shows the result for $K = 9$, beta = 9.0 and indicates a resemblance of the structure to the visual cortex layers.

```
visualize.domain(this_hmrf, 9, 9.0, dot_size=45, size_factor=10,
outfile="visualize.beta.%.1f.png" % 9.0)
```

13. To check if the detected spatial domains are significant, we compare it to a case where the spatial positions of the cells are fully shuffled (or randomly permuted). We first create a randomly permuted data set by shuffling the cells in the original matrix. The parameter 0.99 in the instance method shuffle() is the shuffling proportion (*see* **Note 7**).

```
print "Running pertubed HMRF..."
outdir = "perturbed.jul20"
if not os.path.isdir(outdir):
 os.mkdir(outdir)
perturbed_dset = new_dset.shuffle(0.99)
```

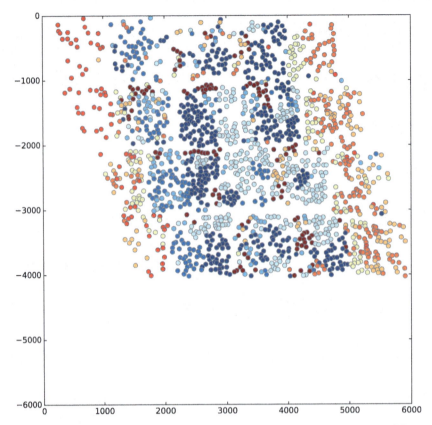

**Fig. 4** The spatial domain organization of the visual cortex, revealed by our HMRF modeling of the spatial data. Each color indicates cells belonging to a spatial domain. There are nine spatial domains ($K = 9$) and beta is set to 9.0

This returns a new data set of DatasetMatrix class. We initiate a new HMRF instance using the same parameters as the original case, and run HMRF on this instance.

```
perturbed_hmrf = HMRFInstance("cortex", outdir, perturbed_dset,
9, (0, 0.5, 30), tolerance=1e-20)
perturbed_hmrf.init(nstart=1000, seed=-1)
perturbed_hmrf.run()
```

14. We compare the log-likelihood of the model in the random case and original case.

```
k=9
betas = np.array(range(0, 90, 5) + range(90, 150, 10)) / 10.0
lik_data, diff_data = [], []
for b in betas:
  lik_data.append((b, "observed", this_hmrf.likelihood[(k,b)]))
  lik_data.append((b, "random",
```

```
perturbed_hmrf.likelihood[(k,b)]))
diff_data.append((b, "obs - rand",
this_hmrf.likelihood[(k,b)] - \
    perturbed_hmrf.likelihood[(k, b)]))
a_lik = pd.DataFrame(data={"label":[v[1] for v in lik_data],
"beta":[v[0] for v in lik_data], "log-likelihood":[v[2] for v in
lik_data]})
d_lik = pd.DataFrame(data={"label":[v[1] for v in diff_data],
"beta":[v[0] for v in diff_data], "log-likelihood":[v[2] for v in
diff_data]})
axn = sns.lmplot(x="beta", y="log-likelihood", hue="label",
data=a_lik, fit_reg=False)
axn = sns.lmplot(x="beta", y="log-likelihood", hue="label",
data=d_lik, fit_reg=False)
```

Here, the output shows the log-likelihood of randomized and observed domain patterns independently (Fig. 5). Log-likelihood is a measure of model fitting and is a function of the number of parameters in the model, beta, among others. The log-likelihood generally increases as beta increases, showing that the model tries to aggressively smooth the domain state by using the neighboring cell states. As log likelihood by itself is less interpretable, we compare it to the shuffled case, and check which beta is the difference largest. The difference between observed and random cases is largest at around beta = 6.0 to 9.0 (Fig. 5). This result likely suggests the following phenomenon: at low beta, the model predominantly uses the cell's own expression to determine the cell state, and is no different from K-means. As beta reaches a critical point,

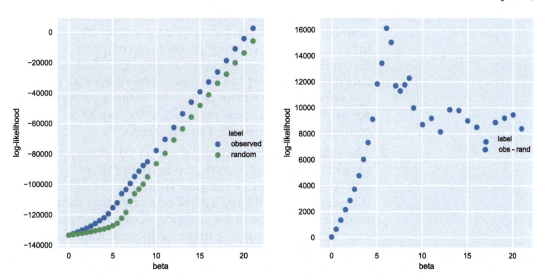

**Fig. 5** Spatial perturbation analysis. We fully shuffled the spatial positions such that 100% of cells' positions are exchanged. (**a**) Log-likelihood of the HMRF model is calculated and plotted for each value of beta from 0 to 20.0. (**b**) Difference in log-likelihood between observed and randomly shuffled data is plotted

neighboring cells collectively influence the cell sate to favor a coherent domain-like pattern, thereby significantly increasing the probability compared to the random case (which is unable to form domains at this critical point due to the lack of spatial relationships). Overall, this critical point may serve as a selection guideline for beta.

15. We provide the entire pipeline as a script for download (see file test_jul20.py at https://bitbucket.org/qzhudfci/smfishhmrf-py). To further facilitate usage and streamline the process, we allow users to define a settings file, which would include input and output files and directories, and various HMRF settings. This settings file is to be used with the pipeline script to obviate the need to modify the script for a new data set (see smfishHmrf-py manual for details). Other example data sets illustrating HMRF are provided on our website http://spatial.rc.fas.harvard.edu (*see* **Note 8**).

## 4    Notes

1. Output:

```
cutoff:0.30% #nodes:1597 #edges:3852 avg.nei:4.82
cutoff:0.50% #nodes:1597 #edges:6384 avg.nei:7.99
cutoff:1.00% #nodes:1597 #edges:12745 avg.nei:15.96
cutoff:2.00% #nodes:1597 #edges:25489 avg.nei:31.92
cutoff:5.00% #nodes:1597 #edges:63721 avg.nei:79.80
```

2. Getting the independent regions of the neighborhood graph is required for the parameter estimation step. This s so that neighboring nodes are not updated one after the other. We turn to the graph coloring problem in computer science to determine the color of the nodes in the graph where neighbors have different colors [20]. The node colors determine the node update order in the EM procedure. By finding equivalent nodes across different independent regions, the aim is to improve consistency of estimates and parallelization.

3. The spatial coherence scoring function calc_silhouette_per_gene has a parameter examine_top to control the proportion of cells (0–1.0) on which to measure coherence. By default, this is 0.1, meaning that the top 10% of cells expressing the gene are used for spatial coherence calculation. It can be increased if the user thinks the spatial pattern is located within a larger proportion.

4. Silhouette output:

```
gene sil.score p-value
amigo2 0.0555237 0.0
cldn5 0.0262171 0.0
calb1 0.0189745 0.0
kcnip 0.0187823 0.0
tbr1 0.0173751 0.0
pax6 0.0169212 0.0
nes 0.0156901 0.0
gda 0.0150629 0.0
col5a1 0.0148561 0.0
loxl1 0.0120843 0.0
sox2 0.011049 0.0
slc5a7 0.00993408 0.0
nov 0.00985005 0.0
itpr2 0.00915686 0.0
cpne5 0.00913211 0.0
Nell1 0.00875134 0.0
mrc1 0.00864791 0.0
rhob 0.00830748 0.0
acta2 0.00802404 0.0
...
Foxa1 0.000510314 0.28
Zfp715 0.000449031 0.34
Galnt3 0.000186029 0.49
Blzf1 -0.000113328 0.64
Laptm5 -0.000492621 0.81
Gm6377 -0.000751389 0.87
Zfp90 -0.00092005 0.91
```

5. Tasic et al. [18] is a complementary scRNAseq data set (GSE71585) which defined 8 major cell types in the visual cortex, including astrocytes, microglia, endothelial cells, three oligodendrocyte clusters, GABA-ergic, and glutamatergic neurons. We use this data set to determine cell-type-specific genes to be removed from HMRF gene list. The goal is to remove these genes from HMRF they are already explained by their cell-type variation. For each gene that is overlapping with our seqFISH data set (i.e., 125 genes) and which is differentially expressed (DE) across cell types, we calculate the gene's average expression z-scores per cell type, as shown in the heat-map (Fig. 6) for 43 DE genes. Star indicates strongly DE genes (avg. expr z-score $> 2.0$), and are selected to be removed from HMRF.

6. The output directory contains many files organized by the beta value, such as the probability estimates, the converged centroids, and the covariance matrices. The file *.prob.txt is a matrix of domain probabilities where each row is a cell and

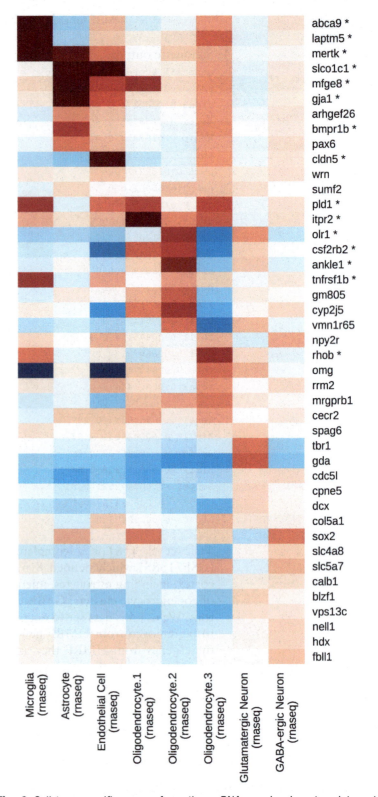

**Fig. 6** Cell-type-specific genes from the scRNAseq visual cortex data set. Average expressions for 44 cell-type-specific genes that were selected from MAST [21] are plotted for each cell type. Star indicates strongly differentially expressed genes (avg. expr z-score > 2.0) that were removed from HMRF

each column is a domain. The assigned domain for each cell is the domain with highest probability value of each row. Results are divided among directories based on K, then within directory by beta.

7. The shuffling proportion (0–1.0) is the proportion of cells that swap positions. For example, 0.99 means that the shuffling rate is such that 99% of cells exchanged position compared to the original data. A low shuffling proportion is good for testing the robustness against spatial noise. When shuffle_prop is set close to 1, all spatial correlations would be destroyed in the data and this is a useful random control case.

8. This website contains all the input files illustrated in this chapter, and an online interactive visualization portal that permits exploration of the parameters of HMRF and their effect on the HMRF model.

## References

1. Deng Q, Ramsköld D, Reinius B, Sandberg R (2014) Single-cell RNA-seq reveals dynamic, random monoallelic gene expression in mammalian cells. Science 343:193–196. https://doi.org/10.1126/science.1245316

2. Jaitin DA, Kenigsberg E, Keren-Shaul H et al (2014) Massively parallel single cell RNA-Seq for marker-free decomposition of tissues into cell types. Science 343:776–779. https://doi.org/10.1126/science.1247651.Massively

3. Macosko EZ, Basu A, Regev A et al (2015) Highly parallel genome-wide expression profiling of individual cells using Nanoliter droplets. Cell 161:1202–1214. https://doi.org/10.1016/j.cell.2015.05.002

4. Schiffenbauer YS, Kalma Y, Trubniykov E et al (2011) Droplet barcoding for single-cell transcriptomics applied to embryonic stem cells. Cell 161:1187–1201. https://doi.org/10.1016/j.cell.2015.04.044

5. Halpern KB, Shenhav R, Matcovitch-Natan O et al (2017) Single-cell spatial reconstruction reveals global division of labour in the mammalian liver. Nature 542:1–5. https://doi.org/10.1038/nature21065

6. Lein ES, Hawrylycz MJ, Ao N et al (2006) Genome-wide atlas of gene expression in the adult mouse brain. Nature 445:168–176. https://doi.org/10.1038/nature05453

7. Raj A, van den Bogaard P, Rifkin SA et al (2008) Imaging individual mRNA molecules using multiple singly labeled probes. Nat Methods 5:877–879. https://doi.org/10.1038/nmeth.1253

8. Lubeck E, Cai L (2012) Single-cell systems biology by super-resolution imaging and combinatorial labeling. Nat Methods 9:743–748. https://doi.org/10.1038/nmeth.2069

9. Chen KH, Boettiger AN, Moffitt JR et al (2015) Spatially resolved, highly multiplexed RNA profiling in single cells. Science 348. https://doi.org/10.1126/science.aaa6090

10. Moffitt JR, Hao J, Bambah-Mukku D et al (2016) High-performance multiplexed fluorescence in situ hybridization in culture and tissue with matrix imprinting and clearing. Proc Natl Acad Sci 113:14456–14461. https://doi.org/10.1073/pnas.1617699113

11. Shah S, Lubeck E, Zhou W, Cai L (2016) In situ transcription profiling of single cells reveals spatial organization of cells in the mouse Hippocampus. Neuron 92:342–357. https://doi.org/10.1016/j.neuron.2016.10.001

12. Zhu Q, Shah S, Dries R et al (2018) Identification of spatially associated subpopulations by combining scRNAseq and sequential fluorescence in situ hybridization data. Nat Biotechnol. https://doi.org/10.1038/nbt.4260

13. Ståhl PL, Salmén F, Vickovic S et al (2016) Visualization and analysis of gene expression in tissue sections by spatial transcriptomics. Science 353:78–82. https://doi.org/10.1126/science.aaf2403

14. Wang Q (2012) HMRF-EM-image: implementation of the hidden markov random field model and its expectation-maximization algorithm. arXiv Prepr

15. Li SZ (2009) Markov random field modeling in image analysis

16. Li SZ (2003) Modeling image analysis problems using Markov random fields. Stoch Process Model Simul 473

17. Obayashi T, Kinoshita K (2011) COXPRESdb: a database to compare gene coexpression in seven model animals. Nucleic Acids Res 39: D1016–D1022

18. Tasic B, Menon V, Nguyen TN et al (2016) Adult mouse cortical cell taxonomy revealed by single cell transcriptomics. Nat Neurosci 19:335–346. https://doi.org/10.1038/nn.4216

19. Chung NC, Storey JD (2015) Statistical significance of variables driving systematic variation in high-dimensional data. Bioinformatics 31:545–554. https://doi.org/10.1093/bioinformatics/btu674

20. Brélaz D (1979) New methods to color the vertices of a graph. Commun ACM 22:251–256. https://doi.org/10.1145/359094.359101

21. Finak G, McDavid A, Yajima M et al (2015) MAST: a flexible statistical framework for assessing transcriptional changes and characterizing heterogeneity in single-cell RNA sequencing data. Genome Biol 16:278. https://doi.org/10.1186/s13059-015-0844-5

# INDEX

Guo-Cheng Yuan (ed.), *Computational Methods for Single-Cell Data Analysis*, Methods in Molecular Biology, vol. 1935,
https://doi.org/10.1007/978-1-4939-9057-3, © Springer Science+Business Media, LLC, part of Springer Nature 2019

Printed by Printforce, United Kingdom